The Curtain: The federal virus flow chart rests on barbed wires in the empty Kansas prarie, superimposed smoke stacks illustrate the flow chart's genocidal intent.

"If an earthquake has two epicenters, like the AIDS virus, rest assured; just as it was in the Wizard of OZ there is someone behind a curtain!"

-Boyd E. Graves

STATE ORIGIN is a gripping review of the direct and shocking government EVIDENCE OF THE LABORATORY BIRTH OF AIDS. Author Boyd Ed Graves has spent the last nine years tracking the origins of AIDS through a densely-covered propaganda trail of the highest deception. His 1999 Flow Chart discovery has been authenticated by medical and scientific doctors around the world. The Flow Chart has been called the greatest document find in the history of mankind.

The shocking conclusions of Graves' research led him to file the landmark federal case Boyd E. Graves vs. The President of the United States, et al. U.S. Supreme Court Case no. 00-9587. His new book "State Origin: The Evidence of the Laboratory Birth of AIDS" should be required reading for the world to ensure we establish a firm foundation, to never again allow the greatest democracy in the history of the world another opportunity to masquerade.

Author Graves provides the irrefutable "missing link," the 1971 AIDS research logic Flow Chart. It is the linkage of over 20,000 scientific papers which conclusively prove a "candidate" virus was developed and mass produced. The "creation", "production" and "proliferation" of AIDS is the ultimate assault on a social fabric woven in democracy. Here, Boyd Ed Graves shares some of the court documents and correspondence in response to the world's most deadly ethnic cleansing experiment, AIDS. The AIDS holocaust has an activist, it is Boyd Ed Graves and this is his story.

STATE ORIGIN:

The Evidence of the Laboratory Birth of

AIDS

A Shocking Collection of Evidence and Court Documents
from Graves vs The President of the United States

U.S. Supreme Court Case No. 00-9587

Boyd Ed Graves, J.D.

First Edition

STATE ORIGIN: THE EVIDENCE OF THE LABORATORY BIRTH OF AIDS

A Shocking Collection of Evidence and Court Documents

Boyd E. Graves v. The President of the United States Supreme Court Case No. 00-9587

First Edition

Published by: National Organization for the Advancement of Humanity & Zygote Media
PO Box 332
Abilene, KS 67410
Phone: 1-800-257-9387
Fax: 1-785-263-1568
email: noah@boydgraves.com
website: http://www.boydgraves.com

First Printing 2001

PRINTED IN THE UNITED STATES OF AMERICA

Library of Congress Cataloging-in-Publication Data Pending
Graves, Boyd E. (July 10, 1952 -)

STATE ORIGIN: The EVIDENCE of the LABORATORY BIRTH of AIDS;
A Shocking Collection of Evidence and Court Documents:
Boyd E. Graves v. The President of the United States, Supreme Court Case No. 00-9587
/ by Boyd Ed Graves - -

1st ed.
p.cm.

Includes biographical references.
ISBN: 0-9707735-1-X

1. AIDS (Disease) - Origins. 2. AIDS (Disease) - Etiology.
3. United States Foreign Policy - Population Control (P.L. 91-213)
3. Human Rights - Eugenics 4. Biological Warfare - Genocide
5. NIH/NCI - Special Virus Cancer Program (1962-1978)
6. Special Virus Cancer Program - 1971 Research Logic Flow Chart
7. United States District Court Ohio - Boyd E. Graves v. Cohen et. al. Case No. 98- 2209
8. Sixth Circuit Court - Graves v. The President of the United States Case No. 99-4476
9. U.S. Supreme Court - Graves v. The President of the United States Case No. 00-9587
9. Biography - Boyd Ed Graves, J.D. (1952 -)

To Mom and Dad

Table of Contents

PHASE I: SELECTION OF SPECIMENS

EXHIBIT 1 — 1.20.2000 Letter, Lisa Hammond Johnson, US Attorney designation of the Office of the President of the United States of America as "true defendant".

EXHIBIT 2 — 5.15.2000 Letter, Victoria Cargill, M.D. M.S.C.E. Medical Officer, Office of AIDS Research, National Institutes of Health.(Official Verification of Special Virus program).

EXHIBIT 3 — August 1972 Flow Chart (US Special Virus program), page 77, Progress Report #9 (Fold-In Chart inside back cover to be used in conjunction with Table III, pages 108-122.) (THE COORDINATION OF EVERY EXPERIMENT IN THE DEVELOPMENT OF AIDS.)

EXHIBIT 4 — 6.9.69 Department of Defense (notification to Congress) testimony seeking to "create" a SYNTHETIC BIOLOGICAL AGENT, pg. 129, Monthly Catalog of United States Government Publications, January 1970, Number 900, Part VI Funding, Chemical and Biological Warfare.

EXHIBIT 5 — FUNDING HISTORY FOR THE DEVELOPMENT OF AIDS (1964 - 1978) pg. 5, Progress Report #15 (1978) of the U.S. Special Virus program. (Note: "Emphasis" grants on pg. 288 (included) reveal present day AIDS biology).

EXHIBIT 6 — Public Law 91-213, The U.S. Population Commission, 3/16/70, Richard M. Nixon, President of the United States of America, (official remarks included).

EXHIBIT 7 — President Nixon's Special Message on Population, July 18, 1969.

EXHIBIT 8 — (undocketed) Petition For Rehearing En Banc (RECEIVED by the Sixth Circuit Court of Appeals, January 29, 2001).

EXHIBIT 9 — Progress Report #9, U.S. Special Virus program, pp. 108 - 122 (1972) ALPHA BETICAL LISTING OF CONTRACTS AND RELEVANCE TO FLOW CHART (pp. 81- 105) (These sections prove beyond an absolute certainty we have the ability to "dismantle" the AIDS virus).

EXHIBIT 10 — National Security Defense Memorandum #46 ("NSDM #46"), written by Zbigniev Brezinski, 3/17/78, "Black Africa and the U.S. Black Movement".

PHASE II: CORRESPONDENCE

Dr. Victoria Cargill, NIH Medical Officer, May 15, 2000
Stephanie Tubbs-Jones, US Congresswoman, December 15, 1998
President William J. Clinton, U.S. President, February 15, 1999
James Clyburn, U.S. Representative, January 5, 1999
John Mangels, Editor Cleveland Plain Dealer, January 8, 1999
John Mangels, Editor Cleveland Plain Dealer, January 20, 1999
Phillip Boffey, Editor- New York Times, January 30, 1999
A.H. Passarella, U.S. Department of Defense, February 9, 1999
Dr. Robert C. Gallo, U.S. Special Virus Project Officer, March 13, 1999
Dr. Micheal Lederman, Case Western Reserve AIDS Unit, May 25, 1999
Mike Dewine, U.S. Senator. May 25, 1999
George Voinovich, U.S. Senator, August 31, 1999
U.S. Senator Ted Kennedy, U.S. Congress, September 8, 1999
Robert G. Anderson, Producer 60 Minutes (CBS), September 9, 1999
George Voinovich, U.S. Senator, September 15, 1999
Dr. Peter Duesberg, U.S. Special Virus Developer, September 15, 1999
Stephanie Tubbs Jones, US Congresswoman, December 12, 1999
Lisa Hammond Johnson, U.S. Department of Justice, January 20, 2000
Dennis J. Kucinich, US Congressman, February 3, 2000
Dr. Victoria Cargill, NIH Medical Officer, September 8, 2000
Dr. Victoria Cargill, NIH Medical Officer, October 14, 2000
Dr. Victoria Cargill, NIH Medical Officer, December 11, 2000
Dr. Brian Foley, Los Alamos AIDS Databse Director, June 18, 2000
Dr. Alan Rabson, U.S. Special Virus Deputy Director, December 17, 2000
Steve Koff, Editor - Cleveland Plain Dealer, November 10, 2000
Ms. Lois Jones, Western Reserve Historical Society, November 24, 2000
Dr. Jonathan Morena, Bioethics Chair University of Vigrinia, July 24, 2000
Dr. Norton Zinder, Rockefeller University, November 10, 2000
Dr. Alan Cantwell, AIDS Origin Researcher/Author, August 31, 2000
Dr. Robert Lee, AIDS Origin Researcher/Author, August 31, 2000
Dr. Robert Lee, AIDS Origin Researcher/Author, September 02, 2000
Dr. Alan Cantwell, AIDS Origin Researcher/Author, September 03, 2000
Dr. Vincent Gammil, AIDS Origin Researcher/ Molecular Chemist September 04, 2000
Dr. David M Hillis, Institute for Cellular and Molecular Biology, October 23, 2000
Dr. Len Hayflick, U.S. Mycoplasma Laboratory Stanford University, August 27, 2000
Dr. Robert Lee, AIDS Origin Researcher/Author, November 6, 2000
Prof. Sir Vincent Mbirika, Sovereign Grand Master Knights of Africa, November 10, 2000
LeRoy Whitfield, Senior Editor POZ Magazine, December 5, 2000
Dr. Victoria Harden, Historian National Institute of Health, December 27, 2001
Dr. Garth Nicolson, Special Virus Developer, February 5, 2001
Dr. Garth Nicolson, Special Virus Developer, February 6, 2001
Dr. Garth Nicolson, Special Virus Developer, February 7, 2001
Gerald Rennerts, Pure Blood, Inc., May 10, 2001

PHASE III-A: TRANSCRIPTS

"The Societal Impact of the Special Virus Program of the U.S.A." - August 21, 1999
"Breaking the Color Barrier: Racial Integration at the U.S. Naval Academy." - November 17, 2000

PHASE III-B: PRESS

AIDS BIOENGINEERING ON TRIAL - January 15, 1999
PENTAGON CONFIRMS TESTIMONY OF DR. DONALD MCARTHUR - February 18, 1999
CDC CALLS AND HANGS UP ON BOYD GRAVES - February 25, 1999
BBC INTERVIEWS GRAVES ON AIDS ORIGINS - September 9, 2000
ELECTION DAY ORDER RULES AIDS BIOENGINEERING "FRIVOLOUS" - November 9, 2000
KANSAS ORGANIZATION SELECTS BOYD E. GRAVES "PERSON OF THE CENTURY" - December 4, 2000
SECRET REPORT REVEALS U.S. SPENT $550 MILLION TO MAKE AIDS - December 6, 2000
GRAVES TO BRIEF PENTAGON ON 1971 AIDS FLOW CHART - January 10, 2001
INTERNATIONAL AIDS ORIGIN ACTIVISTS CONFER ON MLK DAY - January 16, 2001
GRAVES MOVING U.S. TO REVIEW A "SPECIAL VIRUS" PROGRAM - January 31, 2001
WORLD EXPERT JOINS GRAVES CALL FOR AIDS REVIEW - February 9, 2001
GALLO CONFIRMS ROLE IN SPECIAL VIRUS PROGRAM - February 11, 2001
TOM POPE HOSTS AIDS ACTIVIST BOYD GRAVES -February 26, 2001
NAVAL HISTORICAL CENTER TO CHRONICLE ACADEMY LIFE OF BOYD ED GRAVES - March 15. 2001
GRAVES ALERTS U.S. SUPREME COURT TO AIDS BRIEF - March 19, 2001
BOYD E. GRAVES FILES CHARGES WITH U.S. SUPREME COURT - April 11, 2001
DR. BOYD GRAVES TO PRESENT 1971 AIDS FLOWCHART TO UNITED NATIONS RACISM COMMITTEE - April 27, 2001

PHASE IV: COURT DOCUMENTS

CIVIL DOCKET GRAVES v. COHEN, et al 1:98cv2209 U.S. DISTRICT COURT NORTHERN DIVISION
PLAINTIFF'S MOTION FOR RECONSIDERATION ORDER DATED - OCTOBER 28, 1998
CLASS MOTION FOR RECONSIDERATION BASED ON NEWLY DISCOVERED EVIDENCE - September 27, 1999
APPEAL FROM THE FINAL ORDER ENTERED ON 10/27/99 - January 20, 2000
US DEPARTMENT OF JUSTICE NAMING PRESIDENT OF THE US, LISA HAMMOND JOHNSON - January 20, 2000
NOTICE OF ELECTION DAY AIDS ORDER - November 7, 2000
SIXTH CIRCUIT ELECTION DAY AIDS ORDER - November 7, 2000
MOTION FOR RECONSIDERATION OF NOVEMBER 7, 2000 ORDER BASED ON JUDICIAL ERROR - November 11, 2000
JUDGES MERITT, WELLFORD, SILER DENY AIDS PETITION - January 12, 2001
CLASS PETITION FOR REHEARING EN BANC - January 22, 2001
RE: 98-2209 GRAVES v. COHEN (RUMSFELD) SIXTH CIRCUIT NO. 99-4476 - January 27, 2001
U.S. SUPREME COURT CASE NO. 00-9587 CLASS PETITION FOR GLOBAL AIDS APOLOGY - April 11, 2001
U.S. SOLICITOR GENERAL WAIVES U.S. GOVERNMENT'S RIGHT TO RESPOND - May 10, 2001
SUPPLEMENTAL BRIEF IN OPPOSITION TO THE UNITED STATES WAIVER APPLICATION - May 17, 2001

<u>PHASE V: CLINICAL TRAILS</u>

Signature Petition for Immediate Review

Additional Research Materials

One Minute

ACKNOWLEDGMENTS

There have been many people who have bolstered me along the way in so very many ways. I am very fortunate to have had the mom and dad that I did. I have a huge thriving family and a host of friends, classmates and acquaintances all across the country. I am most grateful to those early pioneers of AIDS origin research, particularly Zears Miles, Jr, Ted Strecker and Douglas Huff. I walk in the footsteps of Robert Strecker, Alan Cantwell, Len Horowitz, Robert E. Lee and Dean Loren. I admire the research fortitude of Angie, Sheila, Maureen, Vincent, Tim, George, Elizabeth, Janet, Terry, Eric, Tom, Jason and Basil. I have deep respect and admiration for anyone who may have come in contact with me through this critical journey in search of truth and fact. This road has been long. We appreciate the support from Jon Everett of the Cleveland Life Magazine and Tanya Potts of the Crusader Urban News. Over the course of this journey I have learned my Naval Academy brotherhood was never very far behind. Thank you Ron, Thank you Kerwin, Thank you Bill. Hopefully we can 'beat army' again. Lastly, something must be said of the person who directed this book project, Joel Bales. Thank you Joel for following your vision and your dream. The world may never again want the power of two. If the world would stop tinkering with ethnic bio weapons, we would be willing to consider to dismantle the force. However, I sincerely hope I can continue to rely on the many great people who have helped make this book project a reality. Your strength will continue to guide me throughout the ensuing legal court case and Congressional and U.N. hearings.

Thank you from the bottom of my heart.

Freedom's Slave.

“Blessed is he that readeth.”

Jesus Christ

PREFACE

HISTORY ON HOLD

It was Veteran's Day 2000, four days after the American Presidential election and history was still on hold. Headlines reminded the nation of a divided and hypnotized country preoccupied with the peculiarities of a presidential election in limbo. No headline of the Sixth Circuit's Election Day order refusing Boyd Ed Graves and class the review and apology demanded in their landmark struggle for justice and truth.

I was on my way to Cleveland to meet the man I had only known formally through correspondence. He would call me in Kansas ringing from the Halls of Congress to give me the play by play in America's battle for review of the secret AIDS program he helped uncover, the Special Virus. In 1999, Dr. Graves discovered the program's research logic Flow Chart. Dr. Graves' Flow Chart discovery, was the only story in an international free press for days. I wanted to read more.

Dr. Graves would soon teach me about retro-virus experiments with African Green Monkeys and Human T-Cell Leukemias well underway in these secretly funded laboratories the decades before and leading to the AIDS pandemic. Dr. Graves' document collection, I would later learn, was not even supposed to exist, according to our government.

"Major Findings: An RNA-dependent DNA polymerase similar to that associated with RNA tumor viruses was detected in human leukemic cells but not in normal cells . . .The enzyme was isolated, purified and concentrated 200-fold . . ."

"What are human tissues doing in a monkey experiment?" Dr Graves asks. The evidence spoke for itself, but Dr. Graves made the evidence speak with precision and life. He made the contractual evidence pave the road for international program review. He made the scientific evidence cry for justice.

I wanted to read more. The report year was 1971. The contract was SVCP PR#8 NIH #71-2025, and the leading program scientist, the most famous AIDS doctor in the world, Dr. Robert Gallo. Dr. Graves provided additional documentation freely at his expense, as he had many times to members of Congress, the President, the United Nations, doctors and scientists around the world, and to everyday folk like myself.

Dr. Graves' Flow Chart was as he had maintained, 'The tip of a biological iceberg headed toward an African American Titanic.'

We are all on board.

After over two years in the federal court system, the Sixth Circuit issued a historic order November 7, 2000. Under the cover of a mysterious presidential election, the court ordered Graves' case and the true origin of AIDS - "frivolous." No comment on the Special Virus, or its' Research Logic Flow Chart. Dr. Graves was called 'delusional.'

Working within the framework of the constitutional democracy, Dr. Graves fairly exhausted the process of the judicial system. Without fanfare history was made in Cincinnati, while the nation's eyes were tuned to electoral inequities in Florida. The people were denied in the 2000 election, but the road to justice had only just begun.

I boarded the plane to meet an African American Veteran on a lonely mission to save his race from extinction. On Veteran's Day I came face to face with the international war for review. I met the world leader heading the war, an honest man and the friend who would illustrate humanity's collective challenge and illuminate my personal path to help my fellow brothers and sisters prepare for the most important missions of our lives. Boot Camp is over!

Thank you Dr. Graves.

Joel Bales

National Organization for the Advancement of Humanity

"THE WORLD WAR WITH THE AIDS WEAPON CONTINUES TO RAGE. ONLY KNOWLEDGE AND TRUTH WILL DEFEAT IT."

Boyd E. Graves, J.D. November 2000

Dancing Diffractions

All the time I was underneath, I did not feel fear.
I was conscious and remember it as being tranquil.
The river's current was swift and the sunlight's
diffractions were dancing in a sea of brown.
I was drowning.

"My brother will save me."

"My brother will save me." I kept thinking with certainty.

And then I felt his grasp. I knew my brother was going to save me.

It was not until I reached the surface did I
learn I was in the arms of a white man.

I went into shock. My brother and others
were off in the distance. I was eight years old,
and almost did not see another day.

We were on one of our typical visits to West Virginia
to visit my mom's relatives. They were from an old
bunch of coal miners, who had a fancy for big dinners
with good food.

There was always hearty laughter, which I subsequently learned was
embellished by clandestine drinking.

It was one of those lazy, hazy summer days,
and with my brother and I in tow, my cousin took us
on a meandering stroll.

We bought some ice cream cones and decided to
sit on a railroad tie, looking out over the Kanawha River.

I knew the water was powerful and I didn't know how to swim.
I had no intention of testing it.

Suddenly I was in the water, underneath the water, and moving rapidly.

It had been a prank, a white kid decided to push the smallest Negro kid in the water.

I never saw it coming.

One moment I was admiring the power of the river, the next moment I was in it, and in trouble.

I was completely unaware of the actions of the man at the boat launch downstream.

Our story made the headlines that summer day. Of course when we arrived back at my cousin's house, and the news spread, my brother and he "received discipline."

I was spared, probably because I was dripping wet.

I was very happy to be alive. I never had a chance to hook up with that chicken shit who pushed me in.

I never had an opportunity to express my gratitude to my personal hero, without whose efforts this book would have had no inception.

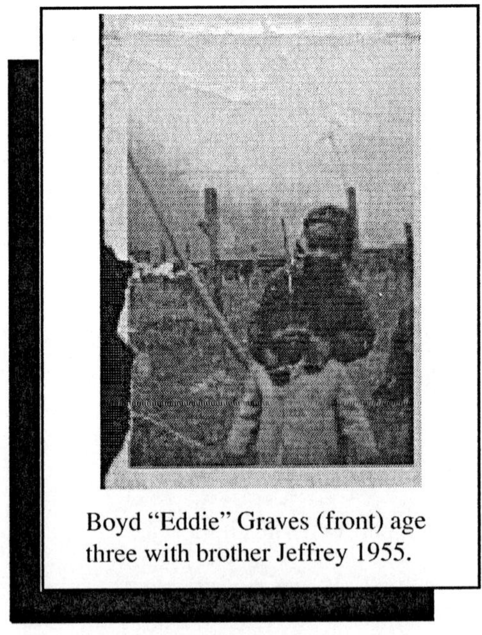

Boyd "Eddie" Graves (front) age three with brother Jeffrey 1955.

TAP & STEP

Several years later I found myself similarly situated.

I was again among friends, but this time there would be no hero to save me.

I had taken upon the task of lead scout, and started out onto the frozen lake. It was mid-December and we were actually on a church retreat. Since some of the guys were scouts, we decided to explore the area around the remote cabin that had been selected by the Sunday school.

We were probably no more than a mile from the cabin, when we discovered the lake. I think initially we were mesmerized by the fact that we had come upon it and it looked frozen. However, no sooner than we arrived I was starting out on a solo trek to 'test the waters'.

With stick in hand, I would tap the ice . . . and step.

The rest of the guys took off around the perimeter. I was all alone. It was a typical, gray Ohio day. There were no reflections from the snow clad surface.

Tap and step.

Tap and step.

I was oblivious to the others. I was concentrating so intently, and I did not realize, I had made it more than halfway.

And then I heard it. The hollow cry of thin ice. To that point, each tap before had only made a thud; my cue to quickly advance to the exact spot I had tapped.

But now I was trapped.

There was no way I could turn around.

There was no way I could go back.

I stepped, and this time I plunged through.

It all happened in an instant, and I immediately felt the icy chill of the cold water below. My friends had made it around to the other side, and heard and saw me plunge in. However, they were powerless to assist.

As they made attempts to come out onto the ice, they quickly found that the lake was not going to allow it. I was submerged in the hole that I had created and was not visible for a number of minutes.

My friends thought I was a goner, I was sure.

But, because I had gone past the center, I found myself able to stand on my tiptoes and stick my nose out of the water. Remembering my scout training, I tried to leap out onto the ice and lay out on the ice like an airplane.

Each time I tried, I again broke through the ice. Eventually, I joined my anxious friends and we began the wet, freezing run back to the cabin.

In the end I just looked stupid. But somewhere deep inside, I was glad I had gone out on the ice. I knew I was going to make it through. I was the lead scout and that was my job.

It is a trait that I have virtually maintained all my life.

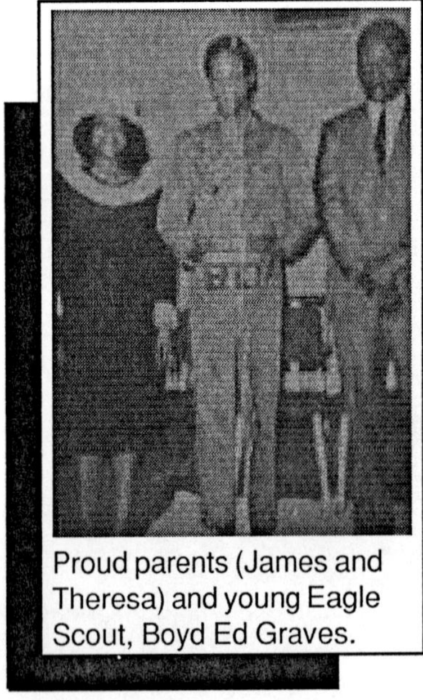

Proud parents (James and Theresa) and young Eagle Scout, Boyd Ed Graves.

EDEN

I was born on July 10, 1952, in Charleston, West Virginia. One month later, my parents and two older brother moved to Youngstown, Ohio and we initially lived with my dad's cousin, Sally. I remember the day we got our own house. It was huge. And so was the back yard. It was like our own little garden of Eden.

There was a grapevine that transgressed the entire fence from the back door to the garage. There were also plum, apple and cherry trees. There were many a year that it was impossible to wait until the fruit ripened before we ate it.

The Joseph family lived next door and they had fruit trees too. Their son, Bruce, and I became best friends, although our parents found more reasons to not embrace.

I would soon learn it was no mistake that my mom and dad had moved into such a big house. I went from being the baby, to being the older brother of six other kids!

The garden of Eden transformed itself into a wasteland of childhood activity. After all, we needed some place to play marbles and kick ball and jump rope. As I look back over the years I am grateful that I am the product of a large family. I have plenty of people to call whenever I need input.

With two older brothers I expectantly only got hand me downs. This was especially obvious since one brother was about my size, and I could grow into my other brother's clothes. Today I have no real fancy for fashion or trends.

I genuinely felt that academics was the single most important endeavor I could undertake. I maintained my love of academics throughout high school, and I found a knack for chess and debating. Somewhere I garnered an interest in campaigning and was elected senior class president. But, we were poor.

My dad worked in the steel mill and with the onset of a continually growing family, even took on a hospital orderly job in New Castle, Pennsylvania. I knew if I were ever going to go to college my parents would not be able to afford it. I studied hard, and somehow, I qualified for the United States' Naval Academy.

It was June, 1971, and I was a plebe.

It did not take long to learn that I was unwelcome. I was Black.

There were a few other Blacks and we did a damned good job of forging a spoken and unspoken brotherhood, which continues to this day. For the record, there are also whites who are indeed part of the lifelong brotherhood. For their continued support, I am humbled and thankful. However, many people there did not like me because I was Black.

Many more joined their category amid rumors of my homosexuality. Black and gay at a staunchly conservative, all white service academy in the early seventies, there in may lie my resolve for perseverance.

This perseverance, has taken me to the greatest confrontation of my winding life. Today I am the lead plaintiff in the lawsuit against the government for bioengineering the AIDS virus.

Before I go further, let me say I have come to this position with eight years of research under my belt. More importantly, I come to this position with direct, credible evidence. The United States' government can not further deny the existence of page 129 of U.S. House Bill 15020/ Page 129 contains the testimony of Doctor Donald MacArthur of the United States Pentagon. On July 1, 1969, Dr. MacArthur outlined to a Congressional subcommittee the necessity for a 'synthetic biological agent.'

Congress awarded the Pentagon the requested funding, and ten years later, consistent with the Pentagon's time frame, we experienced the "big bang." Indeed the gun smoke of Congressional testimony has given rise to the greatest man made tragedy of all time. There is and always has been a diabolical purpose to the AIDS virus, as is verified by the testimony.

You may have personal cause to not like me, and you may have cause to not like the factual message. However nobody can escape the chilling reality of the direct evidence of the laboratory birth of AIDS.

In the fall of 1969, Wolf Szmuness immigrated to the United States from the Soviet Union. By 1978, under the Pentagon's direction, Dr. Szmuness was placed in charge of the Center for Disease Control's Hepatitis B vaccine trials. The Pentagon contaminated the hepatitis B vaccines with their 'synthetic biological agent' and in early 1979, this country had its first case of AIDS.

Dr. Szmuness recruited 1083 promiscuous homosexuals from Manhattan for the initial phase of the trials. In early 1979, Dr. Szmuness effected part two of the Pentagon's plan and began the hepatitis B vaccine trials in other cities with significant homosexual populations. The CDC subsequently documented initial cases of AIDS stemming directly from the Hepatitis B vaccine trials.

There is little doubt of the connection between the Pentagon and the AIDS virus. The Pentagon has perpetrated anarchy, death and destruction for self serving bioengineering projects in the name of military preparedness. A recent front page article suggest the Pentagon has positioned itself to begin a new wave of gene splicing research.

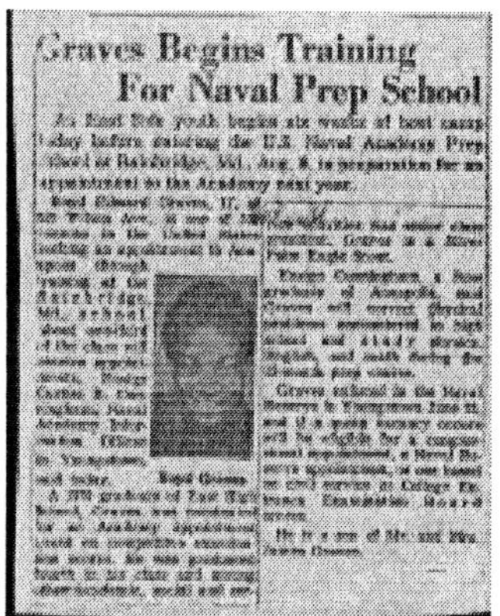

Graves Begins Training For Naval Prep School

At the age of 17, "Eddie" joined the U.S. Navy in search of the Naval Academy in Annapolis, MD. He was the only midshipman to serve the entire year on the Brigade Staff. He would later make history as the first African American elected NAVY GLEE CLUB President, and the first African American pictured on a U.S. Navy recruitment poster. Serving as a communications officer aboard the USS Buchanan Ed was responsible for the nuclear missles launch codes. Ed's reputation for excellence, service, and honor has been documented by historians and honored by the US Navy Historical Center.

"We must let nature determine the finish line, not man. . .We are greater than any federal virus. We are the Human Race."

Boyd E. Graves, J.D.
November 19, 2000

AIDS
THE ATOMIC BOMB OF RACISM

Does it strike a chord that AIDS was made in a laboratory? Does it ring a bell? Does it matter to you that the virus was "recombined" to have an affinity toward a blood marker indigenous to people of color?

Between March 1976 and June 1977, the United States produced 60,000 liters of AIDS. They were ready. They were going to depopulate the Black population by every means necessary. No one could stop them and they made sure of it. (KKK-Kennedy, King, Kennedy).

In 1957, future U.S. President, Gerald Ford participated in the 'off the record' Appropriations hearing that secured funding for a "full scale" offensive biological attack on the Black population. The complete text of that hearing is still classified more than 40 years later! There have been no demands from any of the Black leadership calling for release of the full testimony. Following that hearing, the U.S. had a 'green light' to begin "Special Operation-X" (eXterminate).

On November 6, 1961, ahead of the Special Virus program, the U..S. gave Phizer, Inc. the 'go ahead' to produce "large scale cancer viruses".

THEY ALREADY KNEW VIRUSES CAUSE CANCER!

However without fanfare, the 1957 program was expanded into the Special Virus program in 1962. The program would be a perfect masquerade for the "racial-cleansing agenda". According to the program's history, they were seeking to create, produce and proliferate a 'leukemia/lymphoma' virus.

Do your homework. If you do, you will find that HIV was originally called 'leukemia/lymphoma' virus.

If we review the Flow Chart and 15 progress reports of the federal virus, we will garner immediate medical breakthroughs for people living with HIV/AIDS. The eugenics movement could no longer hold back its desire to execute a scheme to the detriment of the existence of the Black population.

With truth and fact on our side, we can construct a solution for the effective deactivation of AIDS. We have worked our way through the maze and found the "curtain of AIDS". Our future and our national security will be strengthened by our resolve to address this danger in science.

Their greed and self-indulgence clearly show they have a lack of respect for you and human life, as we know it.

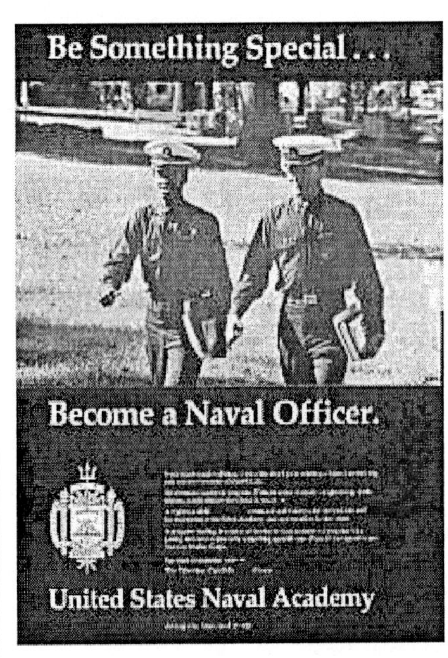

Be Something Special

Boyd "Eddie" Graves made history in 1972 as one of the first African American midshipmen to be pictured on a U.S. Navy Academy recruitment poster. He was also the first African American to be elected president of the internationally acclaimed Navy Glee Club. Dr. Graves' human and civil rights activism has led the world to an irreversible process of accountability and greater equality for all.

NIXON AND THE SPECIAL VIRUS

Few people can understand the depth of the perpetration on behalf of the United States. On one hand we seek to protect (from killing) a fetus. On the other hand, we secretly try to kill the person, particularly if they are Black or socially disenfranchised. Thus in its simplest form, if there is a top secret federal plan or policy to enact covert genocide, somebody would eventually find it.

I first read about the Special Virus in "Emerging Viruses: AIDS & Ebola, Nature, Accident or Intentional?" by Leonard G. Horowitz. I followed Dr. Horowitz's suggestion and ordered progress report #8 (1971) through the public library. About a week later, I was in the library and one of the counter persons said they had something for me. The progress report was about 350 pages. The cover was faded with a very strange design. Dr. Horowitz had accurately described the cover and in bulk form reprinted a number of the pages of the program. As I examined the report for the first time, it appeared to have a blueprint or schematic attached. The flowchart of the program unfolded in five sections and definitively proclaimed this to be the "research logic" of the Special Virus program. I knew it then and I know it now, this flowchart is the "missing link" in proving the laboratory birth of the AIDS pandemic.

In August 1999, I was invited to make a presentation before an international medical research conference in Canada. I had my presentation fairly well in memory and I was glad I did. Midway through my presentation I unveiled the flowchart. A very esteemed scientist was sitting directly in front of me, and nearly fell out of his chair. He demanded to immediately see the flowchart. Shortly after that he had another bombshell to drop. He interrupted again and said, "My name might be in there (the progress reports)!"

Needless, I did return from the conference and ascertained the scientists's role in the development of the Special Virus. Since that time, I have spoken to several of the doctors and scientists who were involved in this secret federal virus development program. It is the people's position that the 'Etiology Area' of the National Cancer Institute (NCI) hold sufficient files to reconstruct important experiments and contracts that will further demonstrate the bioengineering aspect of AIDS. The people also believe the files of Dr. Jon B. Moloney will be crucial to conclusively proving the maintenance and upkeep of this federal virus development program. Dr. Moloney of the NCI was the Chairman of the Special Virus program.

When Dr. Moloney entered into the November 1972 Memorandum of Understanding with the Soviet Union, it was clear the two countries were going to concentrate their ethnic weapon research on the Black population. Although the United States or the Soviet Union would never face an "ethnic oriented attack" from the Black population, they did not have to justify their research. They all secretly wanted to know the difference in the blood of Black and White people.

On April 4, 1969, President Nixon landed by helicopter at Fort Detrick, Maryland. He wanted to be there when our scientists announced to the world that they had figured out the mechanisms of life. We could now create an ethnic cancer. The U.S. could now "enter and control" foreign DNA in the human body. On June 9, 1969 the Pentagon informed the U.S. Congress. On July 18, 1969 Nixon informed the Congress. The United States wanted to take immediate action to thwart overpopulation in the third world. OVer the next nine months Nixon garnered bipartisan support and on March 16, 1970 signed a public law authorizing 'population stabilization'. HIs comments in naming John D. Rockefeller III were perhaps his only truthful statements:

"Of all the people in this nation, I think I could say of all the people in the world, there is perhaps no man who has been more closely identified and longer identified with this problem than John Rockefeller. We are very fortunate to have his chairmanship of the Commission; and we know that the report he will give, the recommendations that he will make, will be tremendously significant as we deal with this highly explosive problem, explosive in every way, as we enter the last third of the 20th century. . ."

Weekly Compilation of Presidential Documents, Vol. VI, page 724, 3/16/70. As enacted the bill (S.2701) is Public Law 91-213 (84 Stat. 67)

The progress reports of the Special Virus program; the experiments and contracts conclusively prove the United States' development of a contagious, biological agent was well underway, prior to the notification the U.S. Pentagon gave the U.S. Congress on June 9, 1969. The U.S. government "feigned confusion" during the late '70's and early '80's solely for the purposes of allowing the special virus to incubate.

The Special Virus program's research logic flow chart and progress reports demonstrate the government's search for a 'candidate virus' and it's ultimate ability to create the AIDS virus. It is necessary for the medical and scientific community to scrutinize the secret memo concerning the "urgent" population problem, former President Nixon issued on July 18, 1969, as well as the plethora of direct scientific evidence affiliated with the program's flowchart and progress reports. Any cursory review of the Special Virus program will conclude the United States developed a human/monkey/sheep retrovirus.

A review of the archives of the National Cancer Institute reveals some of the progress reports of the Special Virus program are located in Bethesda, Maryland. It is worthy to note that some of the progress reports can be accessed through inter-library loan department of the public library system. It is intensely peculiar why our medical and scientific community has been slow to review the experiments and contracts that isolated, developed and proliferated the leukemia/lymphoma virus (AIDS), as is represented by the 15 progress reports of the program. AIDS was constructed in accordance with the experiments and contracts of the Special Virus programs of the United States and others. Specifically, Progress Report #8 (1971) at pages 273-289 contain the specific experiments and contracts that isolated a human retrovirus and co-mingled it with money and sheep viruses. The close homology of HIV and the sheep virus, VISNA is so significant that VISNA virus models are being used for "in vivo" testing of anti-HIV drugs.

Additionally , Progress Report #14 (1977) pages 154-156 contain an experiment involving African children that, if conducted in the United States would be cause for grave civil and human rights concerns. However, Africa has always been the target of a stealth world plan of depopulation sought by world leaders in accordance with the 1974 population conference held in Bucharest Romania.

With respect to the development of the HIV enzyme, Dr. Robert Gallo (and others) isolated a "human" RNA-Dependent DNA Polymerase in the Special Virus program in accordance with a primate inoculation program conducted by Dr. Robert A. Manaker and Dr. Paul A. Kotin. Dr. Gallo's 1971 paper, "Reverse Transcriptase of Type-C Virus Particles of HUman Origin" is remarkably similar to his 1985 "co-discovery" of AIDS!

The scientific connection between the Special Virus and the experimental hepatitis B vaccine lot#:751 from Merck Laboratories is also undeniably evident. The AIDS virus had two epicenters (Manhattan and Africa) because it was added to major vaccine programs in those regions of the world consistent with the Pentagon sworn testimony of June 9, 1969 (McCarthur testimony at 121, Appropriations Hearing Ninety First U.S. Congress June 9, 1969). This information is consistent with an incriminating 1972 scientific paper from the World Health Organization.

The progress reports and the flowchart of the Special Virus program are the "missing links" in the indisputable evidence of the laboratory origin of a 'new' 'chimera' virus. The AIDS virus would not exist but for the pathogenic roots supported by the current "unprofessional silence" of the science and medical communities, goose-stepping for favorable peer reviews. In accordance with Phase IV of the program's flowchart, the United States was able to demonstrate control of the Special Virus (AIDS), prior to proliferation (Phase V).

The medical and scientific communities now have a responsibility to the people of the world to scrutinize the detailed experiments and contracts of the Special Virus program for the secrets to taking it apart and ending its reign of death on Africa, and the world.

In the Spring of 1970 Boyd "Eddie" Graves graduated East High School with honors in academics, sports, and leadership. "Eddie" would soon enter the United States Naval Academy and serve as Communication Officer in the United States Navy during the Nixon administrations. In 1970 President Richard Nixon signed public law PL91-213 authorizing international "population stabilization" policies and legalizing the use of the US Special "AIDS" Virus.

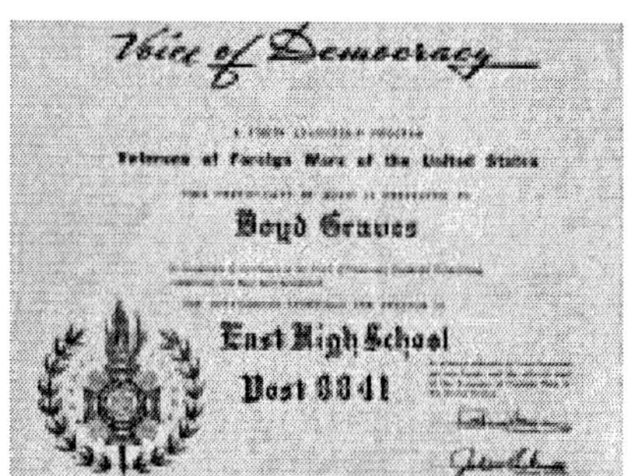

Voice of Democracy

Veterans of Foreign Wars of the United States

Boyd Graves

East High School

Post 8841

"Voice of Democracy" youth leadership award presented to Boyd Graves at Youngstown East High School. Boyd Graves' reputation and skill for leadership has began as a young boy.

"The HIV virus is the result of a century-long hunt for a contagious cancer that selectively kills."

Boyd Ed Graves
POZ Magazine December 2000

CLAY FIGUIRINE

Strecker, Cantwell and Graves. It would turn out to be membership of which I sometimes feel, that maybe there is something to this concept of predispositional existence. The fight for reparations for people with HIV/AIDS begins with me. I am by my own assessment, nothing more than a clay figurine. One subject to weaknesses of personal drought and gluttonous behavior. I openly and honestly submit that I acquired the government's synthetic biological agent during the course of sex.

However, I honestly submit the American homosexual, of which I am also one, was chosen by the purveyors of the experimental vaccine for hepatitis B. It is undisputed that there was an outburst of an unconventional pathogen, in Manhattan, in 1979. That outburst coincided with the start of the Center for Disease Control's (CDC's) Hepatitis B vaccine trials, at the New York Blood Center. Only homosexuals were recruited for the vaccine trials! Although hepatitis B cripples others as well.

This out burst of an unconventional pathogene was, in the government's opinion, a naturally occurring aftershock of the CDC's discovery of unconventional pathogens in Zaire. There is simply no logically occurring exponential projection, that would satisfy the immense number of mathematical improbabilities in which to accept the government's position. It is simply impossible that, whatever the unconventional pathogen, it would not go from a case one in Zaire, in 1976 to a "community-based" budding epidemic in 1979.

It goes further, almost without saying, that this proliferation would be even more difficult, since the alleged discovery of AIDS in Zaire was in a very remote area, not an urban center. Please bear in mind that any time you have a (injecting) virus that strikes a certain, identifiable group, as it was in this case, homosexuals; it should always raise the red flag.

Viruses are not picky. The AIDS virus has most graphically proven this point. Although AIDS was injected into the gay community via the three shots vaccine for hepatitis B, and it was deemed a gay disease. It is now 'gang-busting' its way through the African American heterosexual community. We were foolish to believe that this was a gay virus. Homosexuals have no inherent genetic characteristics that would make us more susceptible to anything!

There is a continuous need for public education.

There is sufficient evidence to suggest the proliferation of the AIDS virus involved the Soviet Union. Wolf Szmuness was the medical doctor who oversaw the CDC's hepatitis B vaccine trials. Wolf Szmuness was the medical doctor who chose homosexuals as a control group.

As an aside, African American sharecroppers were selected as a 'control group' with respect to the infamous Tuskegee syphilis experiments. Quite frankly, I firmly believe the subjects of both control groups have suffered greatly. It is from this perspective that you present the 'correct footing' for assessing the similarities and differences between the groups.

Wolf Szmuness was educated as a medical doctor in Russia. He was a Polish citizen, who at one point shared a rest home room with (a future) Pope John Paul! Dr. Szmuness, after exile to Siberia, was allowed to complete his medical education in the Soviet Union.

At risk for defection, Dr. Szmuness was somehow allowed to travel to Italy with his entire family, where they all defected to the United States. It was the fall of 1969. Several months prior, the government had gotten approval to create a new infectious virus. Now it had a top soviet epidemiologist to run its program.

Less than ten years later, Dr. Szmuness had risen to a tenured professorship at Columbia University and Director of the New York Blood Center's Hepatitis B vaccine trials. Dr. Szmuness was allowed to visit the Soviet Union and return to the United States. If Dr. Szmuness was really a defector, he would have been apprehended by Soviet authorities. He was not.

Dr. Szmuness was sent to the United States at the request of the Department of Defense. At the height of the Cold War it was virtually impossible for a Soviet, with a proclivity for defection, to be allowed to travel outside the Soviet Union. In the few instances when that occurred, the individual would be made to travel alone. By holding back his wife and family, it was a compelling persuasion for the individual to return to the Soviet Union and not defect.

No official explanation explains why Szmuness was allowed out of the Soviet Union with his family. Szmuness had been requested and was now being sent. Thus it was no surprise when Szmuness was not arrested by Soviet officials when he subsequently returned to Russia. The evidence suggests there was a unique relationship between the governments of the United States and Russia with respect to the services of Dr. Wolf Szmuness.

I honestly believe Dr. Szmuness chose homosexuals as the control group because he believed the virus would not jump into other communities. Dr. Szmuness was naively deceived. Any virus allegedly capable of jumping species easily skips through communities. In order to set up the pandemic explosion, it was necessary to prep the fuse with homophobic and religious derangement.

Much of which continues to this day, despite the fact that African American women are the latest identifiable group dying en masse. It would not be uncommon to hear a current Black pastor decry the unholiness and absolute sins of people who love differently than they. It is the hypocrisy of stone throwers. Oh how easy it is to ridicule.

If there were a modern day Jesus, I'm sure he would belong to an AIDS service organization. I'm sure he would accent and highlight the love and compassion, the more difficult path. Be weary of people who readily represent hatred and intolerance in the name of some supreme being. Often times they are merely espousing interpersonal issues reflective of their own guilt, shame or confusions.

Nonetheless, the U.S. Government did seek to create a synthetic biological agent and Dr. Szmuness was contracted to proliferate it. Szmuness chose gay Americans to spread the biological agent. It is undisputed that the Center for Disease Control had the AIDS virus in its possession in 1976. Two years later, the government distributed its hepatitis B vaccine to homosexuals only. It is clear the CDC came out of the bush in Zaire in 1976, with the AIDS virus in tow. It is very peculiar that AIDS and ebola viruses both emerged from the same remote area of Zaire, seemingly at about the same time.

Prior to 1976, there are no stored blood samples of HIV positive blood.

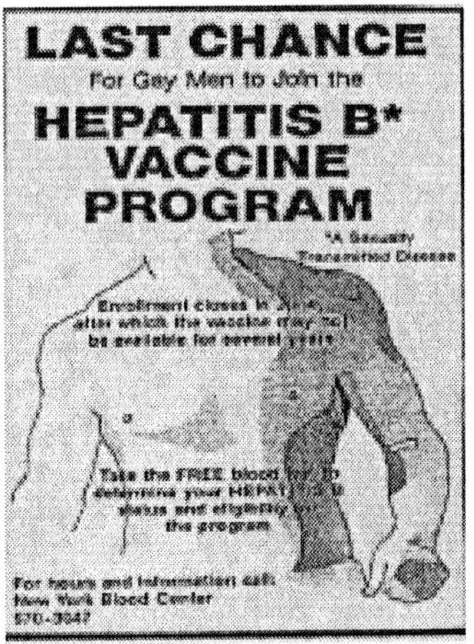

"LAST CHANCE For Gay Men to Join the HEPATITIS B* VACCINE PROGRAM **a sexually transmitted disease*" reads the title to the New York Blood Center's 1979 poster recruiting promiscous homosexuals for innocluation under clinical trails (See Flow Chart Phase-V). Blood tests later revealed the new vaccine successfully immunized the gay volunteers against the Hepatitis B virus; however, within several short years over 66% of these previously healthy young men were dead from a mysterious new **sexually transmitted disease* called G.R.I.D. (Gay Related Immune Deficiency). Unkowingly, they spread the new virus to others, and became the first innocent American victims of a special virus we know today as HIV/AIDS.

THE HISTORY OF THE DEVELOPMENT OF AIDS

The true history of the origin of AIDS can be traced throughout the 20th Century and back to 1878. On April 29 of that year they United States passed a "FEDERAL QUARANTINE ACT".

The United States began a significant effort to investigate "causes" of epidemic diseases. In 1887, the effort was enhanced with the mandate of the U.S. "LABORATORY OF HYGIENE". This lab was run by Dr. Joseph J. Kinyoun, a deep rooted-racist, who served the eugenics movement with dedication.

Two years later, 1889, we were able to identify "mycoplasmas", a transmissible agent, that is now found at the heart of human diseases, including (AIDS) HIV.

In 1893, we strengthened the Federal Quarantine Act and suddenly there was an explosion of polio.

In 1898, we knew we could use mycoplasma to cause epidemics, because we were able to do so in cattle, and we saw it in tobacco plants.

In 1899, the U.S. Congress began investigating "leprosy in the United States".

In 1902, We organized a "Station for Experimental Evolution" and we were able to identify diseases of an ethnic nature.

In 1904, we used mycoplasma to cause an epidemic in horses.

In 1910, we used mycoplasma to cause an epidemic in fowl/birds.

In 1917, we formed the "Federation of the American Society for Experimental Biology" (FASEB).

In 1918, the influenza virus killed millions of unsuspecting. It was a flu virus modified with a bird mycoplasma for which human primates had no "acquired immunity."

In 1921, lead eugenics philosopher, Betrand Russell, publicly supported the "necessity for "organized" plagues" against the Black population.

In 1931, we secretly tested African Americans and we tested AIDS in sheep.

In 1935, we learned we could crystallize the tobacco mycoplasma, and it would remain infectious.

In 1943, we officially began our bio-warfare program. Shortly thereafter, we were finding our way to New Guinea to study mycoplasma in humans.

In 1945, we witnessed the greatest influx of foreign scientists in history into the U.S. biological program. Operation Paperclip will live in infamy as one of the darkest programs of a twisted parallel government fixated on genocide.

In 1946, the United States Navy hired Dr. Earl Traub, a notorious racist biologist.

A May appropriations hearing confirms the existence of a "secret" biological weapon.

In 1948, we know that the United States confirmed the endorsement of "devising a scheme" in which to address the issue of overpopulation in certain racial groups. State Department's George McKennan's memo will forever illuminate the eugenics mendacity necessary for genocide of millions of innocent people.

In 1949, Dr. Bjorn Sigurdsson isolates the visna virus. Visna is man made and shares some "unique DNA" with HIV. See, Proceedings of the United States, NAS, Vol. 92, pp. 3283 - 7, (April 11, 1995).

In 1951, we now know our government conducted its first virus attack on African Americans. Crates in Pennsylvania were tainted to see how many Negro crate handlers in Virginia would acquire the placebo virus.. They were also experimentally infecting sheep and goats. According to author Eva Snead, they also held their first world conference on an AIDS-like virus.

In 1954, Dr. Bjorn Sigurdsson publishes his first paper on Visna virus and establishes himself as the "Grandfather of the AIDS virus." He will encounter competition from Dr. Carlton Gajdusek.

In 1955, they were able to artificially assemble the tobacco mosaic virus. Mycoplasmas will forever be at the heart of the U.S. biological warfare program

In 1957, future U.S. president, Rep Gerald Ford and others gave the U.S. Pentagon permission to aggressively deploy offensive biological agents. There are no recorded cases of AIDS prior to the 1957 creation of "Special Operation-X." (The SOX) program served as the immediate prototype program for the Special Virus program to begin in 1962.

By 1960, Nikita Kruschev had been let in on the biological weapon. His 1960 statement will long reflect the arrogance of the secret blend of communism and democracy. The two countries would go to a November 1972 agreement to cull the Black Population.

In 1961, scientist Haldor Thomar publishes that viruses cause cancer. In 1995, he and Carlton Gajdusek informed the National Academy of Sciences that "the study of visna in sheep would be the best test for candidate anti-HIV drugs."

In 1962, under the cover of cancer research, the United States charts a path to commit premeditated murder, the "Special Virus" program begins on February 12th. Dr. Len Hayflick sets up a U.S. mycoplasma laboratory at Stanford University. Many believe the "Special Virus" program began in November 1961 with a Phizer contract.

Beginning in 1963 and for every year thereafter, the "Special Virus" program conducted annual progress reviews at Hershey Medical Center, Hershey, PA. The annual meetings are representative of the aggressive nature in which the United States pursued the development of AIDS.

In 1964, the United States Congress gave full support for the leukemia/ lymphoma (AIDS) virus research.

In 1967, the National Academy of Sciences launched a full scale assault on Africa. The CIA (Technical Services Division) acknowledged its secret inoculator program.

In 1969, Fort Detrick told world scientists and the Pentagon asked for more money, they knew they could make AIDS. Nixon's July 18 secret memo to Congress on "Overpopulation" serves as the start of the paper trail of the AIDS Holocaust.

In 1970, President Nixon signed PL91-213 and John D. Rockefeller, III became the "Population Czar." Nixon's August 10 National Security Memo leaves no doubt as to the genocidal nature of depopulation.

In 1971, Progress Report #8 is issued. The flowchart (pg. 61) will forever resolve the true laboratory birth origin of AIDS. Eventually the Special Virus program will issue 15 reports and over 20,000 scientific papers. The flowchart links every scientific paper, medical experiment and U.S. contract. The flowchart would remain "missing" until 1999. World scientists were stunned.

The flowchart will gain in significance throughout the 21st Century. It is also clear the experiments conducted under Phase IV-A of the flowchart are our best route to better therapy and treatment for people living with HIV/AIDS. The first sixty pages of progress report #8 of the Special Virus program prove conclusively the specific goal of the program. By June 1977, the Special Virus program had produced 15, 000 gallons of AIDS. The AIDS virus was attached as complement to vaccines sent to Africa and Manhattan. However, because of the thoroughness of authors, like Dr. Robert E. Lee, we also learn the Stanford Mycoplasma Laboratory issues one of the first papers with AIDS in the title. "Viral Infections in Man Associated with Acquired Immunological Deficiency States." The primary scientist, Dr. Thomas Merigan, was a "consultant" to the Special Virus program.

Progress Report # 8 at 104 - 106 proves Dr. Robert Gallo was secretly working on the development of AIDS with full support of the sector of the U.S. government that seeks to kill its citizens. Dr. Gallo can not explain why he excluded his role as a "project officer" for the Special Virus program from his biographical book. Dr. Gallo's early work and discoveries will finally be viewed in relation to the flowchart. We now know where every experiment fits into the flowchart. The "research logic" is irrefutable evidence of a federal other undesirables. Tomorrow it may be you.

INAUGURATION DAY 1973

January 20, 1973 Boyd "Eddie" Graves breaks ranks during Nixon's second inauguration to get an autograph from U.S.O. legend Bob Hope. As the Secret Service looks on, the entire Whitehouse photo crew snapped this photo.

(Offical Whitehouse Photo)

Progress Report #8 at 273 - 286 proves we gave AIDS to monkeys. Since 1962, the United States and Dr. Robert Gallo have been inoculating monkeys and re-releasing them back into the wild. Thus, even government scientists are baffled that both HIV-1 and HIV-II would "suddenly emerge" from two distinct monkey ancestral relatives during the last 100 years. A 1999 Japanese study will ultimately prove the Man to Monkey origin of Monkey AIDS. The monkey experiments summary definitively proves Monkey AIDS is also man-made.

In 1972, the United States and the Soviet Union entered into a biological agreement that would signal the death knell for the Black Population. The 1972 agreement for collaboration and cooperation in the development of offensive biological agents is still U. S. policy.

In 1973, we find that world scientist, Garth Nicolson reports on his project, "Role of the Cell Surface in Escape From Immunological Surveillance." His report is accompanied by seven published papers. Dr. Nicolson worked in conjunction with the Special Virus program from 1972 until 1978. Dr. Nicolson is considered by some to be Dr. Gallo's "West Coast" counterpart. It is strongly held that because of Dr. Nicolson, Dr. Robert Gallo and Dr. Luc Montagnier would secretly meet in Southern California to coordinate what they would and would not say about the special virus development program.

In 1974, Furher Henry Kissinger releases his NSSM-200 (U.S. Plan to Address Overpopulation). It is the only issue of discussion at the World Population Conference in Bucharest, Romania. • The men in the shadows had won, the whole world agrees to secretly cull Africa's population. Today it is Africa and other undesirables. Tomorrow it will be you.

In 1975, President Gerald Ford signs National Security Defense Memorandum #314. The United States implements the Kissinger NSSM-200.

In 1976, the United States issues Progress Report #13 of the Special Virus program. The report proves the United States had various international agreements with the Russians, Germans, British, French, Canadians and Japanese. The plot to kill Black people has wide international support. In March, the Special Virus began production of the AIDS virus, by June 1977, the program will have produced 15,000 gallons of AIDS. President Jimmy Carter allows for the continuation of the secret plan to cull the Black Population.

In 1977, Dr. Robert Gallo and the top Soviet Scientists meet to discuss the proliferation of the 15,000 gallons of AIDS. They attach AIDS as complement to the Small pox vaccine for Africa, and the "experimental" hepatitis B vaccine for Manhattan. According to authors June Goodfield and Alan Cantwell, it is Batch #751 that was administered in New York to thousands of innocent people. This government will never be able to repay the people for the social rape, humiliation and out right prejudice people with HIV/AIDS face on a daily basis. The men in the shadows of the AIDS curtain accurately calculated that you would not care if only Blacks and gays are dying. In fact you don't care that nearly a half million Gulf War veterans are encumbered with something contagious. Soon there will be no more Black people and a confused military, older White people will start suddenly dying and you still won't get it. Be here now for us, give us a chance to be there for you.

Suddenly, just as President Nixon had predicted, there was explosive death. On November 4, 1999, the U.S. White House announced,....Within a period as short as five years, all new infections of HIV in the United States will be African American...." At some point our experts must be allowed to begin the interface process of allowing the history of this virus program to count. It is ludicrous and preposterous to fail to review the U.S. virus program in which to elucidate the etiology of AIDS.

More of the history of the secret virus program can be found in the archives of Dr. John B. Maloney. A review of the files under Dr. Moloney's name would further pinpoint additional dates and records consistent with one of the greatest hunts, capture and proliferation of disease in the history of the human race.

We have found the missing link. It is the guts of the research logic of a federal program that seeks to kill. We have found a curtain of AIDS. We can identify some of the people who work in the shadows of the curtain. Dr. Robert Gallo and Dr. Garth Nicolson must lead us in inmediate review. In light of the attack mechanisms available in which to inhibit AIDS, it is time that not another person be stricken with this relic, synthetic mycoplasma chimera.

We are all one people. Help those of us who are still here to realize full and contributory lives. On September 28, 1998 I filed suit against the United States for the "creation", "production" and "proliferation" of AIDS. On November 7, 2000, the appeals court agreed with the lower court and held AIDS bioengineering as "frivolous." The world continues to wait for the court to rule on the resubmitted issues. The court can not continue to simply brush aside our experts and the government's flowchart.

We can not allow the state an autocratic right to govern outside of the Constitution. Our society is structured to hide crimes committed by the state, while punishing citizens for minor indiscretions. Their strategy focuses on the general confusion they can create via manipulation of the media. They are very good at what they do. We must become more focused in our continued presentation of the flowchart. The flowchart is the absolute missing link in proving the existence of a coordinated research program to develop a cancer virus that depletes the immune system. New diseases do not create old illnesses.

I have been asked to give my perspective with regard to the federal program MK-NAOMI . MK-NAOMI is the code for the development of AIDS. The "MK" portion stands for the two co-authors of the AIDS virus, Robert Manaker and Paul Kotin.

The "NAOMI" portion stands for "Negroes are Only Momentary Individuals." The U.S. government continues to orchestrate silence from the very top echelons of the Congress and military. At present there is no accountability. The good people will ultimately create a tsunami of public outrage.

This compilation of court documents and correspondence is the true effort of one man's achievement in solving the mystery of the origin of AIDS. We have found the origin of AIDS, it is our nation state, it is us.

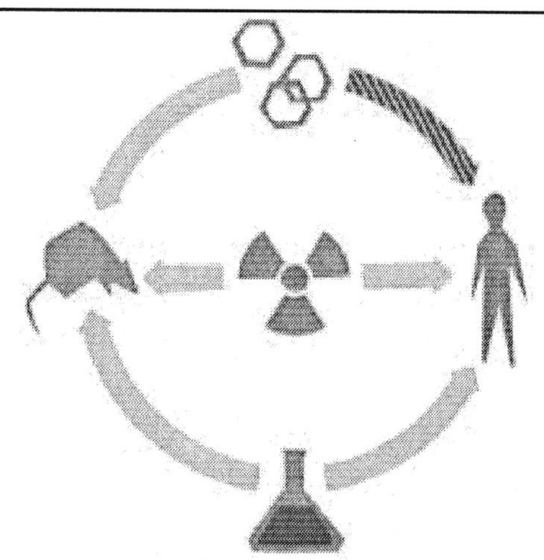

Logo of the U.S. Special Virus Cancer Program (1962-1978).
The logo symbolically relates the program's function of moving lower animal viruses into human hosts cells through viral insertion technlogy and chemical manipulation. Progress report #8 reveals the program had mastered the retrovirus in 1971 and gene splicing research. The present day Human Genome Project headed by Craig Venter's Celera Genomics has worked with the U.S. government to feverishly map the Human genome as well as their second priority genome now near complete . . .the mouse. Top: (three six sided virus particles - mycoplasmas) Center: (radiation). Bottom: (chemical reaction). Left: (mouse) Right: (Human)

"It is no longer a question of IF the TRUTH will be known, but WHEN, and those who have been directly involved, (as well as those who were culpably responsible for indirectly doing nothing to STOP the genocide), will be brought to justice in the World Court & in the eyes of history & humanity, if they have not already had to ingnomiously account to their 'Maker'. God Rest their SOULS."

Basil Earle Wainwright
Physicist Three Time Nobel Prize Nominee.

AIDS: GENOCIDE

OF THE HEART, SOUL, AND MIND.

Thirty years ago, President Richard Nixon signed a public law (91-213) that authorized "population stabilization". None of us have ever read the law. If we were to read, we would find that President Nixon appointed John D. Rockefeller, III to be the "population czar". John Rockefeller had spent his entire life pursuing the racist ideology of "eugenics". In Mr. Rockefeller's world, the Black population had to be significantly reduced (if not exterminated). All the government documents reveal 'the United States was devising a means to effect mass death on the Black population'.

We will end the 20th Century in the midst of the greatest state-sanctioned murder program in the history of the world.

In other words, the flowchart proves the AIDS virus is a man-made, 'designer virus' meant to depopulate the African community. The AIDS virus is the candidate virus sought by the United States in accordance with their 'research logic' identified in the 1971 flowchart of the 'Special Virus' program. The 1971 flowchart is the 'irrefutable missing link' which proves the laboratory birth of AIDS. The 'Special Virus' program of the United States began officially in 1962 and issued 15 yearly progress reports until 1978. The current HIV/aids databases of the United States do not cross reference the databases of the federal virus program, the 'Special Virus', or the databases of the U.S. mycoplasma computers.

In February, I hand-delivered 3,000 signature petitions to Rep. Tubbs Jones in Washington, calling for an immediate review of this federal virus program. Rep. Tubbs Jones refuses to publicly acknowledge the signature petitions or take any action on them!

Complacency is the enemy of the Black population.

However, if the Black population is asleep or comatose, and its (so called) leaders are on the white man's front porch, sooner or later it will not be necessary for Black folk to awaken at all!

We have bought into a 'centralized intelligence' that implores us to reaffirm daily, our religious faith. We ignorantly and blindly transfer our faith in religion, to faith in government. Since 1887, and the creation of the U.S. 'Laboratory of Hygiene', they have worked incessantly to create a 'contagious cancer that selectively kills'. In 1931, while they were setting up blacks in Tuskegee, they were also conducting the first tests of the AIDS virus in the sheep population in Iceland. Further evidence of the sheep connection (Visna) to AIDS is graphically articulated in the Proceedings of the United States (1995).

What does it matter to tell you this, if none of you are going to read? I am sure it is not that you don't need that is the problem, perhaps we stymy ourselves by continuously re-reading one book!

We have the 'direct evidence' of the laboratory birth of AIDS. The Flowchart has exceeded international reviews since August of 1999. The AIDS Flowchart coordinates and links over 20,000 scientific papers and absolutely proves the bioengineering of AIDS.

Today (9/28/00) is the second anniversary of the federal AIDS lawsuit. The case has been ripe for appeal in the Sixth Circuit (Case No. 99-4476) since March, 2000. Surely they are not illegally holding the decision in this case until after the November presidential election. Expert after expert continue to conclude the issue of AIDS bioenginnering is not frivolous, or in contradiction to an apology for the 'creation', 'production' and 'proliferation' of AIDS. We are entitled to an immediate review of the flowchart and the progress reports of the 'Special Virus' program of the United States of America. The United States can not now deny to revelations of Phase IV-A of the flowchart. It does appear the United States has been 'maintaining a cure' to AIDS, all along!

When Nixon appointed Rockefeller to oversee this 'problem', the plan to kill Blacks was only supposed to extend until the end of the year 2000. It is time to 'deactivate' the AIDS Holocaust. I appeal to our inherent resolve to demand immediate review of the government structure existing behind the curtain of AIDS. Similar to the movie, the Wizard of Oz, we have found a government curtain.

Who amongst you will assist our process in seeking to know the truths beyond the curtain? I have seen behind the curtain, it is a lily-white world of unhappiness and frustration.

NOW IS THE TIME to mentally awake and join our international call for an independent review of this federal virus development program.

The people are entitled to a review of this federal virus program. The government has yet explain the whereabouts of 12 of the 15 progress reports! Dr. Alan Rabson (NCI) could not explain why the National Cancer Institute (NCI) Archives has only 3 of the 15 reports.

We will continue to aggressively demand accountability and explanation within the confines of our established national constitution. The Black population and others are entitled to an explanation for the race-cleansing mentality (eugenics) that has dominated the 20th Century, and has left massive death and intellectual rape.

Richard M. Nixon 37th President (1969-1974)

"...of all the people in this nation, I think I could say of all the people in the world, there is perhaps no man who has been more closely identified and longer identified with this problem (population) than John Rockefeller. We are very fortunate to have his chairmanship of the Commission; and we know the report that he will give, the recommendations that he will make, will be tremendously signficant as we deal with this highly explosive problem, explosive in every way, as we enter the last third of the 20th century."

*Weekly compilation of Presidential Documents (vol. 6 p. 734)

The Smoking Gun of AIDS: A 1971 Flowchart

In 1977, a secret federal virus program produced 15,000 gallons of AIDS. The record reveals the United States was represented by Dr. Robert Gallo and the USSR was represented by Dr. Novakhatskiy of the diabolical Ivanosky Institute. On August 21, 1999, the world first saw the flowchart of the plot to thin the Black Population.

The 1971 AIDS flowchart coordinates over 20,000 scientific papers and fifteen years of progress reports of a secret federal virus development program. The epidemiology of AIDS is an identical match to the "research logic" identified in the five section foldout. The flowchart is page 61 of Progress Report #8 (1971) of the Special Virus program of the United States of America. We today, challenge world scientists to discussion of this document find. We believe there is a daily, growing number of world experts who are all coming to the same conclusion regarding the significance of the flowchart. Dr. Garth Nicolson has examined the flowchart as well as other top notch experts from around the world. It is time for Dr. Michael Morrissey of Germany to examine the flowchart and report to the world. In addition, we have now examined the 1978 report. It is heresy to continue to further argue the program ended in 1977. The 1978 report of the development of AIDS leaves no doubt as to the ("narrow result") candidate virus sought by the United States. The flowchart conclusively proves a secret federal plot to develop a "contagious cancer" that "selectively kills."

Following the presentation of the flowchart in Canada, the same information was presented to the United States in the rotunda of the Western Reserve Historical Society in Cleveland. Shortly thereafter a major African newspaper called and for four days in a row, this issue was the feature story in an uncensored press. The people of Africa already know about the U.S. virus development program. It is time for the rest of us to know.

In January, the U.S. had no response to my two page abstract submitted to the African American AIDS 2000 conference. In February, the U.S. Congress had no response to the 3000 Americans who signed signature petitions calling for immediate review of the flowchart and progress reports of the secret virus development program. We firmly believe once the dust settles from the current election marathon, reviewing the special virus program will be the single most important pursuit of the 21st Century.

More scientists and doctors must join with Dr. Nicolson, Dr. Strecker, Dr. Cantwell, Dr. Horowitz, Dr. Lee, Dr. Wainwright, Dr. Halstead and Professor Boyle. In any public debate on this issue, we will continue to present the flowchart of the secret virus development program, as the "irrefutable missing link" in the true laboratory origin of AIDS. We have successfully navigated a federal maze and matrix and found a curtain surrounding the issue of AIDS. The 1999 discovery and presentation of the AIDS flowchart is a "smoke detector" wake up call. Society has an obligation to do more than don masks.

Non-inclusive random endnotes:

U.S. Special Virus program, Progress Report #8 (1971), pg. 61 (the flowchart)

National Security Defense Memorandum (NSDM) #314, Brent Scowcroft (1975).

"Special Message to the U.S. Congress on Problems of Population Growth", Richard M. Nixon, July 18, 1969

Public Law 91-213, "To Stabilize World Populations", John D. Rockefeller, III, Chairman, March 16, 1970

National Security Council Memorandum (NSCM) #46, "Black Africa and the U.S. Black Movement", Zbigniew Brezinski, March 17, 1978

Boyd E. Graves, J.D. made the first formal presentation of the 1971 Special Virus Flow Chart during the First Annual Common Cause Medical Research Foundation Conference held in Canada in 1999. His special abstract on the origin of AIDS, "STATE ORIGIN: The Evidence of the Laboratory Birth of AIDS" was originally written for the AIDS-2000 Conference in Washington D.C. Boyd E. Graves' shocked the medical and scientific world with his presentation of the absolute evidence of a global AIDS genocide program. On January 20, 2000 the United States Department of Justice named the office of the President of the United States to answer Dr. Graves' petition for global AIDS apology and the immediate review and deactivation of the Special Virus Cancer program (1962-1978). The USDOJ's action was in direct response to the submission of the 30 year old AIDS Flow Chart into evidence.

A SPECIAL ABSTRACT

ON THE MEDICAL ETIOLOGY OF AIDS

"STATE ORIGIN:

The Evidence of the Laboratory Birth of AIDS"

by: Boyd E. Graves., B.S., J.D.,

PRESENTER'S OVERVIEW:

This paper is a call for independent medical and scientific review of the Special Virus program of the United States of America. The program's 1971 flowchart does display the 'research logic' of an ultra secret federal program that seeks to create a 'leukemia/lymphoma' virus. In 1975, President Gerald Ford signed into law, "National Security Defense Memorandum #34, ("NSDM#34"). NSDM#34 was official US approval of the enactment of the United States' (stealth) depopulation programs as outlined in National Security Study Memorandum-200. ("NSSM-200"), written by Henry Kissinger in 1974.

The progress reports of the Special Virus program; the experiments and contracts, conclusively prove the United States' development of a contagious, biological agent was well underway, prior to the notification the US Pentagon gave the US Congress on June 9, 1969. It is the presenter's position the US government "feigned confusion" during the late '70's and early '80's solely for the purposes of allowing the special virus to incubate. It is the presenter's position that the AIDS virus can be definitively shown that it was "put together"; we may then be able to demonstrate a renewed vigor in our assault on the virus. It is also necessary for the medical and scientific community to scrutinize the secret memo concerning the "urgent" population problem, former President Nixon issued on July 18, 1969, as well as the plethora of direct scientific evidence affiliated with the programs' flowchart and progress reports. Any cursory review of the Special Virus program will conclude the United States developed a human/monkey/sheep retrovirus.

ABSTRACT

A review of the archives of the National Cancer Institute reveals some of the progress reports of the Special Virus programs are located in Bethesda, Maryland. It is worthy to note that some of the progress reports can be accessed through the interlibrary loan department of the public library system. It is intensely peculiar why our medical and scientific community has been slow to review the experiments and contracts that isolated, developed and proliferated the leukemia/lymphoma virus (AIDS), as is represented by the 15 progress reports of the program. It is the presenter's position that the AIDS virus was constructed in accordance with the experiments and contracts of the Special Virus programs of the United States and others. Specifically, Progress Report #8 (1971) at pages 273 - 289 contain the specific experiments and contracts that isolated a human retrovirus and co-mingled it with monkey and sheep viruses, inter alia. The close homology of HIV and the sheep virus, VISNA, is so significant that VISNA virus models are being used for "in vivo" testing of anti-HIV drugs.

Additionally, Progress Report #14 (1977) at pages 154 - 156 contain an experiment involving African children that, if conducted in the United States, would be cause for grave civil and human rights concerns. However, Africa has always been the target of a stealth world plan of depopulation sought by world leaders in accordance with the 1974 population conference held in Bucharest, Romania. With respect to the development of the HIV enzyme, Dr. Gallo (and others) isolated a "human" RNA-Dependent DNA Polymerase in the Special Virus program in accordance with a primate inoculation program conducted by Dr. Robert A. Manaker and Dr. Paul A. Kotin.

Dr. Gallo's 1971 paper, "Reverse Transcriptase of Type-C Virus Particles of Human Origin" is remarkably similar to his 1984 "co-discovery" of AIDS. The presenter can also show a scientific connection between the Special Virus and the experimental hepatitis B vaccine lot: 751 from Merck Laboratories. The AIDS virus had two epicenters (Manhattan and Africa) because it was added as complement to major vaccine program in those regions of the world consistent with the Pentagon sworn testimony of June 9, 1969, (See MacArthur testimony at 121, Appropriations Hearings, June 9, 1969, Appropriations Subcommittee of the Ninety-first US Congress). This information is also consistent with an incriminating 1972 scientific paper from the World Health Organization.

Finally, the presenter has spoken to all four of the experts on the audiotape, "Virus Makers of The CIA". The flowchart and the progress reports of the special virus are the "missing links" in the indisputable evidence of the laboratory origin of a "new" "chimera" virus. The AIDS virus would not exist but for its iatrogenic roots supported by the current "unprofessional silence" of the science and medical communities, goose-stepping for favorable peer reviews. In accordance with Phase IV of the program's flowchart, the United States was able to demonstrate control of the Special Virus (AIDS), prior to proliferation (Phase V).

It is the presenter's position the medical and scientific communities can now prove the AIDS virus was bioengineered. The medical and scientific communities have a responsibility to the people of the world to scrutinize the detailed experiments and contracts of the Special Virus program for the secrets in taking it apart.

Endnotes: **"A Special Abstract on the Medical Etiology of AIDS."**

1. 1971 AIDS DEVELOPMENT FLOW CHART—U.S. Special Virus Cancer Program, Progress Report #8 (1971), pg. 61, National Cancer Institute, Etiology Area.

2. National Security Defense Memorandum #314 (1975)

3. National Secuirty Defense Memorandum #200, Henry Kissinger (1974).

4. Progress Report #8 at 2 (1971).

5. "Special Message to the U.S. Congress on Problems of Population Growth", Richard Nixon, July 18, 1969.

6. The world cataloging system of the public library lists eight references to the Special Virus program. See also; Baker, C., et.al., "The Special Virus Leukemia Program of the National Cancer Institute", Some Recent Developments in Comparative Medicine", edited by Fiennes (Londn: Academic Press, 1966).

7. Page 276 identifies NIH contract #71-2025, NIH contract #71-2025 is located at pp. 104 - 106 and reveals the "isolation" of a human retrovirus, PR#8 (1971).

8. Thomar, H. et.al. (presented by CARLTON GAJDUSEK), Inhibitory effect of 9-(2-phosphonylmethoxyethyl)adenine on visna virus infection in lambs; a model for in vivotesting of candidate anti-human immunodeficiency virus drugs. Proceedings of the National Academy of Sciences of the United States of America, Volume 92, pp. 3283 - 3287 (April 11, 1995).

9."Sero-Epidemiologic and Laboratory Studies on Nasopharyngeal Carcinoma and Burkitt's Lymphoma", International Agency for Research on Cancer, Progess Report #14, pp. 251 - 253 (1977).

10. Progress Report #8 at 105.

11. Progress Report #8 at 335.

12. Goodfield, J. "Quest for the Killers", (Boston, Massachusetts, Birkhauser, 1985).

13. U.S. Department of Defense Appropriations Hearing, June 9, 1969, U.S. House Resolution 15090, Part VI, pages 121 & 129.

14."Biological Significance of Histocompatibility Antigens", Committee for the Conference, John E. Fogarty Center for Advanced Study in the Health Sciences and the World Health Organization, Federation of American Societies for Experimental Biology, Federation Proceedings, Volume 31, No. 3, May - June 1972. Conference held at the National Institutes of Health, Bethesda, Maryland, July 27 -30, 1970.

15. "Virus Makers of the CIA", Cantwell, Haslam, Horowitz, Nicolson, KFJC-FM, Los Altos, California, DAVID EMORY, One Step Beyond (1997).

notes:

AIDS: A HIDDEN BIOLOGICAL WORLD

On November 7th, the Presidential election provided significant cover in which to hide a critical event concerning the destiny of Black America. The Sixth Circuit Court of Appeals in Cincinnati ruled that the issue of AIDS bioengineering is frivolous. None of us cared. We were too busy to care. To many of us, AIDS is not an issue. We have our sports, we have our movies and music, and we have our religion. We are comfortable. We are comfortable in knowing that AIDS is not currently affecting us and our family, so we don't have to deal with it. The Sixth Circuit accurately predicted that Black America would have no response to its violation of the rule of law of the United States, and they were partially right.

According to Sir Vincent of the Knights of Africa, 'the timing of the ruling was no accident'. The sixth Circuit waited eight months in which to sneak the ruling in under the draft of a tremendously peculiar election. The Sixth Circuit accurately predicted that no one would be concerned about the origin of AIDS, if they coupled it with the frenzy of a national election day. They were wrong.

The good people of the United States and the world are concerned about violations of the rule of law, and we are concerned about the truth of the origin of AIDS. We will continue to fight to assist our fellow brothers and sisters to awaken from the deep hypnotic sleep of mass mind control. The evidence clearly show the AIDS virus did not originate in Africa, it began in the laboratories of our government. They made it. They mass produced it and they complemented vaccines with it. As smallpox ended in Africa, AIDS began. As an experimental vaccine was tested in Manhattan, AIDS began. However, to confuse you, our government has laid the framework for mass hypnosis through vigorous propaganda of an African cesspool of disease. Because you are content, you believe them without them offering any legitimate proof what so ever. Because of religion, sports and music you are completely unaware that your soul has long ago, already been sold down river.

The hunt for a "contagious cancer" goes back to the early part of the century. They have been seeking the separate blood markers for the Black population as far back as 1902. However, this is of no concern to you. You have placed your chips and those of your descendants in your faith, and that alone will see you through.

Unfortunately you are wrong.

You must seek to know more.

Disturb your comatose complacency. Ask Congresswoman Tubbs Jones for more accountability from our government. The "Special Virus" program is real. Our reliance on our faith and our love for sports and music will not save us. They have calculated that we are insensitive to our own destiny.

They are merely appeasing you, until there are none of us left.

WAKE UP! START ASKING QUESTIONS! GET INVOLVED!

With your help, motivation and concern, we can change the direction of the boat. We must work together to save the Black population from a holocaust that has been brewed behind an AIDS curtain. Make no mistake about it, the AIDS curtain is real.

We have successfully navigated the maze and the matrix and found a hidden biological world. We must now collectively demand review of this secret, stealth program seeking to exterminate the Black population.

It is your move.

More than 16 million people have died from AIDS since 1980: three times the number killed during the Jewish Holocaust. 60 percent of the dead are in sub-Saharan Africa. In the U.S., African American's represent over 80% of deaths and new infections.

Source: The Washington Post, World Bank, WHO, UNICEF, USAID (4/20/00)

NEGATIVE EUGENICS:

MACHINERY FOR CONTINUOUS PRODUCTION OF AIDS

Recently I had a very fortunate opportunity to be afforded some insight into the mindset of the late 19th Century as it relates to the forerunners of what subsequently became known as "eugenics". Only a very, tiny amount of people on this planet have ever heard of the "Station for Experimental Evolution'. Check it out for yourself starting about 1902. The point I want to emphasize, is that we know now that it is inherently unfair to "exclude inclusion" in a process (supposedly) destined for all. By this I mean, in theory, we would all agree that we should include input from everyone (or everyone's representatives) and devise a "multi-tasking" intelligence quotient consistent and reflective of our intellectual tomorrows.

We have within our reach, a break in this storm with "tilted" clouds. Level the playing fields of tomorrow today, We should never again allow them to be turned into monolith biological killing fields.

We must devise our tomorrow with input from all. I sincerely believe we can all be made better and I commit to my clay footprints in well trod sand.

I have been truly humbled beyond explanation, yet I remain intellectually defiant.

Eugenics is an inherently flawed principle best illuminated by its narrow genesis (May 1, 1776). THEY have simply set up a 'favoring' intellectual quotient, to ensure you are made aware, according to their standard, you are a peon. It is precisely this tangent which tears out the heart of a true Constitutional form of government and mentally rapes and enslaves true good people for, as we see here, a couple of hundred years. In this rubric of the necessity for biological technology and its ensuing evil, we must all 'trip the switch' of self-realization and internal spiritualization. As we begin to awaken from mental 'suspended animation', we must know that we can catch up, and supercede. They have always known, to play fair, they have to continually tilt the field.

In light of the usages of some of the synthetic microorganisms (e.g., Mycoplasma Visna (AIDS)), ***without our reawakening, the negative slant of eugenics will surely vanquish us.***

PHASE I

SELECTION OF SPECIMENS

www.boydgraves.com

"The U.S. Special Virus program spent $550 million dollars to make a virus and has never accounted for the program to the American people ***until now.****"*

Boyd E. Graves, J.D.
excerpted from the United States Supreme Court Appeal filed April 11, 2001

Supreme Court, U.S.
FILED
APR 1 1 2001
OFFICE OF THE CLERK

March 27, 2001

Clerk of the Court
United States Supreme Court
1 First Street, N. E.
Washington, D.C. 20543

RE: RULE 32 DIAGRAM & EXHIBITS: for inspection: Boyd E. Graves v The President of the United States, Case No. 99-4476, an appeal from the Sixth Circuit

Dear Office of the Clerk:

Pursuant to Rule 32, Petitioner, Boyd E. Graves, submits the enclosed list, following diagram and exhibits in support of his allegations of AIDS bioengineering to be placed in the custody of the clerk for inspection. These documents are submitted in accordance with a timely appeal from the Sixth Circuit Court of Appeals, Case No. 99-4476. I submit these documents as evidence of a burning social issue in need of the High Court's attention. I will formally petition this Honorable Court on April 11, 2001 at 11:00 a.m. Please provide a file stamp copy of this letter. Thank you.

Thank you for unwavering devotion to the American people.

Respectfully submitted,

Boyd E. Graves, Petitioner (Class Representative)
for AIDS Apology and Review of the U.S. Special Virus Program
(Case No. 99-4476-Sixth Circuit)
P.O. Box 332
Abilene, KS 67410
785-263-1871

cc: Solicitor General

DIAGRAM AND EXHIBIT LIST OF PETITIONER, BOYD E. GRAVES PURSUANT TO RULE 32

Boyd E. Graves v The President of the United States, et al. an appeal from the Sixth Circuit Court of Appeals, Cincinnati, Ohio - - Case No. 99-4476 to the United States Supreme Court

---

EXHIBIT 1	1.20.2000 Letter, Lisa Hammond Johnson, US Attorney designation of the Office of the President of the United States of America as "true defendant".
EXHIBIT 2	5.15.2000 Letter, Victoria Cargill, M.D. M.S.C.E. Medical Officer, Office of AIDS Research, National Institutes of Health.(Official Verification of Special Virus program).
EXHIBIT 3	August 1972 Flow Chart (US Special Virus program), page 77, Progress Report #9 (Fold-In Chart inside back cover to be used in conjunction with Table III, pages 108-122.) (THE COORDINATION OF EVERY EXPERIMENT IN THE DEVELOPMENT OF AIDS.)
EXHIBIT 4	6.9.69 Department of Defense (notification to Congress) testimony seeking to "create" a SYNTHETIC BIOLOGICAL AGENT, pg. 129, Monthly Catalog of United States Government Publications, January 1970, Number 900, Part VI Funding, Chemical and Biological Warfare.
EXHIBIT 5	FUNDING HISTORY FOR THE DEVELOPMENT OF AIDS (1964 - 1978) pg. 5, Progress Report #15 (1978) of the U.S. Special Virus program. (Note: "Emphasis" grants on pg. 288 (included) reveal present day AIDS biology).
EXHIBIT 6	Public Law 91-213, The U.S. Population Commission, 3/16/70, Richard M. Nixon, President of the United States of America, (official remarks included).
EXHIBIT 7	President Nixon's Special Message on Population, July 18, 1969.
EXHIBIT 8	(undocketed) Petition For Rehearing En Banc (RECEIVED by the Sixth Circuit Court of Appeals, January 29, 2001).
EXHIBIT 9	Progress Report #9, U.S. Special Virus program, pp. 108 - 122 (1972) ALPHA BETICAL LISTING OF CONTRACTS AND RELEVANCE TO FLOW CHART (pp. 81- 105) (These sections prove beyond an absolute certainty we have the ability to "dismantle" the AIDS virus).
EXHIBIT 10	National Security Defense Memorandum #46 ("NSDM #46"), written by Zbigniev Brezinski, 3/17/78, "Black Africa and the U.S. Black Movement".

Exhibit 1

U.S. Department of Justice

United States Attorney
Northern District of Ohio

1800 Bank One Center
600 Superior Avenue, East
Cleveland, Ohio 44114-2654

January 20, 2000

Mr. Boyd E. Graves
2700 Washington Street
Cleveland, Ohio 44113

Re: *Boyd E. Graves v. The President of the United States, et al.*, Court of Appeals Case No. 99-4476

Dear Mr. Graves:

Enclosed is a copy of the Appearance of Counsel form which was mailed to the court on January 19, 2000.

Very truly yours,

Lisa Hammond Johnson/pdh

Lisa Hammond Johnson
Assistant U.S. Attorney
216/622-3679

Enclosure

pdh

NATIONAL INSTITUTES OF HEALTH
OFFICE OF
AIDS
RESEARCH

Victoria A. Cargill, M.D., M.S.C.E.
Medical Officer

Building 2, Room 4E20
Two Center Drive
MSC 0255
Bethesda, MD 20892
TEL: 301 402 2932
FAX: 301 496 4843
E-Mail: cargillv@od.nih.gov

MAY 15 2000

Boyd E. Graves, Esq.
2700 Washington Street
Apartment 1402
Cleveland, Ohio 44113

Dear Mr. Graves:

I am writing to follow up on our telephone conversation last week regarding the flow chart on the Human Immunodeficiency Virus (HIV). As I indicated on the telephone, I called you in response to a telephone contact you made to Ms. Linda Jackson in our office regarding this flow chart. Using the information you provided, I located and read the materials on your web sites:

http://www.alltheweb.com
http://www.aidsbiowar.com.

Subsequent to our telephone conversation, I located and contacted Mrs. Judith Grossberg, the librarian at the National Cancer Institute archives. She confirmed the presence of the reports on the Special Virus, and obtained report #8. I have reviewed this report, as well as the flow chart, to which you referred.

I have referred this matter, and the report citations to the National Cancer Institute, as the research was conducted within their Institute. I have provided for them the specific report numbers, and citations which you shared with me, as well as your web sites indicated above.

Thank you for your interest in AIDS.

Sincerely,

Victoria A. Cargill, MD. MSCE

Victoria A. Cargill, M.D., M.S.C.E.
Medical Officer

Exhibit 3

1971 Special Virus Flow Chart Fold Out 11x36"

- Here-

http://www.boydgraves.com/flowchart

EBOOK CUSTOMERS PROCEED TO PHASE V

insert foldout here

CONGRESS. HOUSE (Appropriations, Committee on)

EXHIBIT 1

DEPARTMENT OF DEFENSE APPROPRIATIONS FOR 1970

HEARINGS

BEFORE A

SUBCOMMITTEE OF THE COMMITTEE ON APPROPRIATIONS HOUSE OF REPRESENTATIVES

NINETY-FIRST CONGRESS

FIRST SESSION

SUBCOMMITTEE ON DEPARTMENT OF DEFENSE APPROPRIATIONS

GEORGE H. MAHON, Texas, *Chairman*

ROBERT L. F. SIKES, Florida
JAMIE L. WHITTEN, Mississippi
GEORGE W. ANDREWS, Alabama
DANIEL J. FLOOD, Pennsylvania
JOHN M. SLACK, West Virginia
JOSEPH P. ADDABBO, New York
FRANK E. EVANS, Colorado[1]
GLENARD P. LIPSCOMB, California
WILLIAM E. MINSHALL, Ohio
JOHN J. RHODES, Arizona
GLENN R. DAVIS, Wisconsin

R. L. MICHAELS, RALPH PRESTON, JOHN GARRITY, PETER MURPHY, ROBERT NICHOLAS, AND ROBERT FOSTER, *Staff Assistants*

[1] Temporarily assigned.

PART 6

Budget and Financial Management
Budget for Secretarial Activities
Chemical and Biological Warfare
Defense Installations and Procurement
Defense Intelligence Agency
Safeguard Ballistic Missile Defense System
Testimony of Adm. Hyman G. Rickover
Testimony of Members of Congress and Other Individuals and Organizations

Printed for the use of the Committee on Appropriations

Exhibit 1

agents that we have ever considered. So, we have to believe they are probably working in the same areas.

SYNTHETIC BIOLOGICAL AGENTS

There are two things about the biological agent field I would like to mention. One is the possibility of technological surprise. Molecular biology is a field that is advancing very rapidly, and eminent biologists believe that within a period of 5 to 10 years it would be possible to produce a synthetic biological agent, an agent that does not naturally exist and for which no natural immunity could have been acquired.

Mr. SIKES. Are we doing any work in that field?

Dr. MACARTHUR. We are not.

Mr. SIKES. Why not? Lack of money or lack of interest?

Dr. MACARTHUR. Certainly not lack of interest.

Mr. SIKES. Would you provide for our records information on what would be required, what the advantages of such a program would be, the time and the cost involved?

Dr. MACARTHUR. We will be very happy to.

(The information follows:)

The dramatic progress being made in the field of molecular biology led us to investigate the relevance of this field of science to biological warfare. A small group of experts considered this matter and provided the following observations:

1. All biological agents up to the present time are representatives of naturally occurring disease, and are thus known by scientists throughout the world. They are easily available to qualified scientists for research, either for offensive or defensive purposes.

2. Within the next 5 to 10 years, it would probably be possible to make a new infective microorganism which could differ in certain important aspects from any known disease-causing organisms. Most important of these is that it might be refractory to the immunological and therapeutic processes upon which we depend to maintain our relative freedom from infectious disease.

3. A research program to explore the feasibility of this could be completed in approximately 5 years at a total cost of $10 million.

4. It would be very difficult to establish such a program. Molecular biology is a relatively new science. There are not many highly competent scientists in the field, almost all are in university laboratories, and they are generally adequately supported from sources other than DOD. However, it was considered possible to initiate an adequate program through the National Academy of Sciences-National Research Council (NAS-NRC).

The matter was discussed with the NAS-NRC, and tentative plans were made to initiate the program. However, decreasing funds in CB, growing criticism of the CB program, and our reluctance to involve the NAS-NRC in such a controversial endeavor have led us to postpone it for the past 2 years.

It is a highly controversial issue, and there are many who believe such research should not be undertaken lest it lead to yet another method of massive killing of large populations. On the other hand, without the sure scientific knowledge that such a weapon is possible, and an understanding of the ways it could be done, there is little that can be done to devise defensive measures. Should an enemy develop it there is little doubt that this is an important area of potential military technological inferiority in which there is no adequate research program.

Exhibit 5

"$550 million in roughly 'INVISIBLE Appropriations'" to make AIDS

1/15/01

Table 1

FUNDING HISTORY OF THE VIRAL ONCOLOGY PROGRAM
(in thousands)

Fiscal Year	Number of Positions	In-House	VO Contracts	SVLP	VCP	BCTF	CREG	TOTALS
1964	30	—	4926	—	—	—	—	
1965	117	1687	5433	8723	—	—	—	15,843
1966	140	1835	3064	13,556	—	115	—	18,570
1967	144	1999	3137	13,505	—	246	—	18,887
1968	157	2239	—	—	17,241	284	—	19,764
1969	176	2891	—	—	17,985	259	—	21,135
1970	180	3356	—	—	17,340	174	—	20,870
1971	197	4517	—	—	31,591	234	—	36,342
1972	226	6310	—	—	41,889	734	—	48,933
1973	219	6983	—	—	42,564	1006	—	50,553
1974	231	7189	—	—	49,553	1150	—	57,892
1975	229	9395	—	—	49,387	1450	—	60,232
1976	222	10,800	—	—	46,773	1450	967	59,990
1977	234	12,547	—	—	44,450	1450	1800	60,247
1978	234	15,296	—	—	41,171	1450	1800	59,717

gay
+
env
nomenclature

Exhibit <u>6</u>

March 16, 1970. The official remarks of President Richard Nixon upon Signing Public Law 91-213 establishing the Commission on Population Growth and the American Future, chaired by John D. Rockefeller III. A White House release of March 16, 1970 announcing the signing of the bill is printed in the Weekly Compilation of Presidential Documents (vol. 6, p. 734).

Ladies and gentlemen:

We have asked you into this room because the Cabinet Room is presently being redecorated. The purpose is to sign the population message. I shall sign the message and make a brief statement with regard to it.

First, this message is bipartisan in character as is indicated by the Senators and Congressmen who are standing here today. This is the first message on population ever submitted to the Congress and passed by the Congress. It is time for such a message to be submitted and also the time to set up a Population Commission.

Let me indicate very briefly come of the principles behind this population message.

First, it will study both the situation with regard to population growth in the United States and worldwide.

Second, it does not approach the problem from the standpoint of making an arbitrary decision that population will be a certain number and stop there. It approaches the problem in terms of trying to find out what we can expect in the way of population growth, where that population will move, and then how we can properly deal with it.

It also, of course, deals with the problem of excessive population in areas, both in nations and in parts of nations, where there simply are not the resources to sustain adequate life.

I would also add that the Congress, particularly the House of Representatives, I think, contributed very much to this message by adding amendments indicating that the Population Commission should study the problems of the environment as they are affected by population, and also that the Population Commission should take into account the ethical considerations that we all know are involved in a question as sensitive as this.

I believe this is an historic occasion. It has been made historic not simply by the act of the president in signing this measure, but by the fact that it has had bipartisan support and also such broad support in the nation.

An indication of that broad support is that John D. Rockefeller has agreed to serve as Chairman of the Commission. The other member of the Commission will be announced at a later time. Of all the people in this nation, I think I could say of all the people in the world, there is perhaps no man who has been more closely identified and longer identified with this problem that John Rockefeller. We are very fortunate to have his chairmanship of the Commission; and we know that the report that he will give, the recommendations that he will make, will be tremendously significant as we deal with this highly explosive problem, explosive in every way, as we enter the last third of the 20th century.

And I again congratulate the Member of the House and Senate for their bipartisan support. I wish the members of the Commission well.

And as usual we have pens for all the Members of Congress who participated in making this bill possible and for the members of the staff who are present here.

NOTE: The President spoke at 10:16 a.m. in the Roosevelt Room at the White House.

UNITED STATES STATUTES AT LARGE

CONTAINING THE

LAWS AND CONCURRENT RESOLUTIONS
ENACTED DURING THE SECOND SESSION OF THE
NINETY-FIRST CONGRESS
OF THE UNITED STATES OF AMERICA

1970—1971

AND

REORGANIZATION PLANS AND PROCLAMATIONS

VOLUME 84

IN TWO PARTS

PART 1

PUBLIC LAWS 91-191 THROUGH 91-525

UNITED STATES
GOVERNMENT PRINTING OFFICE
WASHINGTON : 1971

(3) not to exceed twenty members appointed by the President.

(b) The President shall designate one of the members to serve as Chairman and one to serve as Vice Chairman of the Commission.

(c) The majority of the members of the Commission shall constitute a quorum, but a lesser number may conduct hearings.

COMPENSATION OF MEMBERS OF THE COMMISSION

SEC. 3. (a) Members of the Commission who are officers or full-time employees of the United States shall serve without compensation in addition to that received for their services as officers or employees of the United States.

(b) Members of the Commission who are not officers or full-time employees of the United States shall each receive $100 per diem when engaged in the actual performance of duties vested in the Commission.

80 Stat. 499; Stat. 190.

(c) All members of the Commission shall be allowed travel expenses, including per diem in lieu of subsistence, as authorized by section 5703 of title 5 of the United States Code for persons in the Government service employed intermittently.

DUTIES OF THE COMMISSION

SEC. 4. The Commission shall conduct an inquiry into the following aspects of population growth in the United States and its foreseeable social consequences:

(1) the probable course of population growth, internal migration, and related demographic developments between now and the year 2000;

(2) the resources in the public sector of the economy that will be required to deal with the anticipated growth in population;

(3) the ways in which population growth may affect the activities of Federal, State, and local government;

(4) the impact of population growth on environmental pollution and on the depletion of natural resources; and

(5) the various means appropriate to the ethical values and principles of this society by which our Nation can achieve a population level properly suited for its environmental, natural resources, and other needs.

STAFF OF THE COMMISSION

80 Stat. 378. 5 USC 101 et q.

80 Stat. 443, 9.

Post, p. 198-1.

SEC. 5. (a) The Commission shall appoint an Executive Director and such other personnel as the Commission deems necessary without regard to the provisions of title 5 of the United States Code governing appointments in the competitive service and shall fix the compensation of such personnel without regard to the provisions of chapter 51 and subtitle III of chapter 53 of such title relating to classification and General Schedule pay rates: *Provided*, That no personnel so appointed shall receive compensation in excess of the rate authorized for GS-18 by section 5332 of such title.

80 Stat. 416.

(b) The Executive Director, with the approval of the Commission, is authorized to obtain services in accordance with the provisions of section 3109 of title 5 of the United States Code, but at rates for individuals not to exceed the per diem equivalent of the rate authorized for GS-18 by section 5332 of such title.

Contract au- ority.

(c) The Commission is authorized to enter into contracts with public agencies, private firms, institutions, and individuals for the conduct of research and surveys, the preparation of reports, and other activities necessary to the discharge of its duties.

(3) Sections 397, 398, 399, 399a, and 399b of such Act are redesignated as sections 396, 397, 398, 399, and 399a, respectively. 79 Stat. 1063. 42 USC 280b-7.

(d)(1) The part of title III of such Act redesignated as part I is amended by striking out "Surgeon General" each place it occurs in the sections of such part redesignated as sections 382, 383, 386, and 388. The section of such part redesignated as section 384 is amended by striking out "Surgeon General" and inserting in lieu thereof "Board". Ante, p. 66.

(2)(A) The part of title III of such Act redesignated as part J is amended by striking out "Surgeon General" each place it occurs and inserting in lieu thereof "Secretary".

(B) The subsection of section 393 of such part redesignated as subsection (e) is amended by striking out "Surgeon General's" and inserting in lieu thereof "Secretary's". Ante, p. 64.

MEANING OF SECRETARY

SEC. 11. Subsection (c) of section 2 of title I of the Public Health Service Act (42 U.S.C. 20) is amended to read as follows:

"(c) Unless the context otherwise requires, the term 'Secretary' means the Secretary of Health, Education, and Welfare." 58 Stat. 682. 42 USC 201 and note.

EFFECTIVE DATE

SEC. 12. (a) Except as provided in subsection (b), the amendments made by this Act shall apply with respect to appropriations for fiscal years ending after June 30, 1970.

(b) The amendments made by sections 10(d) and 11 shall take effect on the date of the enactment of this Act.

Approved March 13, 1970.

Public Law 91-213

AN ACT

To establish a Commission on Population Growth and the American Future.

March 16, 1970 [S. 2701]

Be it enacted by the Senate and House of Representatives of the United States of America in Congress assembled, That the Commission on Population Growth and the American Future is hereby established to conduct and sponsor such studies and research and make such recommendations as may be necessary to provide information and education to all levels of government in the United States, and to our people, regarding a broad range of problems associated with population growth and their implications for America's future.

Commission on Population Growth and the American Future. Establishment.

MEMBERSHIP OF COMMISSION

SEC. 2. (a) The Commission on Population Growth and the American Future (hereinafter referred to as the "Commission") shall be composed of—

(1) two Members of the Senate who shall be members of different political parties and who shall be appointed by the President of the Senate;

(2) two Members of the House of Representatives who shall be members of different political parties and who shall be appointed by the Speaker of the House of Representatives; and

GOVERNMENT AGENCY COOPERATION

SEC. 6. The Commission is authorized to request from any Federal department or agency any information and assistance it deems necessary to carry out its functions; and each such department or agency is authorized to cooperate with the Commission and, to the extent permitted by law, to furnish such information and assistance to the Commission upon request made by the Chairman or any other member when acting as Chairman.

ADMINISTRATIVE SERVICES

SEC. 7. The General Services Administration shall provide administrative services for the Commission on a reimbursable basis.

REPORTS OF COMMISSION; TERMINATION

SEC. 8. In order that the President and the Congress may be kept advised of the progress of its work, the Commission shall, from time to time, report to the President and the Congress such significant findings and recommendations as it deems advisable. The Commission shall submit an interim report to the President and the Congress one year after it is established and shall submit its final report two years after the enactment of this Act. The Commission shall cease to exist sixty days after the date of the submission of its final report.

AUTHORIZATION OF APPROPRIATIONS

SEC. 9. There are hereby authorized to be appropriated, out of any money in the Treasury not otherwise appropriated, such amounts as may be necessary to carry out the provisions of this Act.

Approved March 16, 1970.

Public Law 91-214

AN ACT

To amend Public Law 89-260 to authorize additional funds for the Library of Congress James Madison Memorial Building.

March 16, 1970 [S. 2910]

Be it enacted by the Senate and House of Representatives of the United States of America in Congress assembled, That section 3 of the joint resolution entitled "Joint resolution to authorize the Architect of the Capitol to construct the third Library of Congress building in square 732 in the District of Columbia to be named the James Madison Memorial Building and to contain a Madison Memorial Hall, and for other purposes", approved October 19, 1965 (79 Stat. 986), is amended by striking out "$75,000,000" and inserting in lieu thereof "$90,000,000".

Library of Congress James Madison Memorial Building. Appropriation increase.

2 USC 141 note.

SEC. 2. Nothing contained in the Act of October 19, 1965 (79 Stat. 986), shall be construed to authorize the use of the third Library of Congress building authorized by such Act for general office building purposes.

Prohibition.

Approved March 16, 1970.

Exhibit 7

PRESIDENT RICHARD NIXON'S JULY 18, 1969
"SPECIAL MESSAGE TO CONGRESS ON THE PROBLEMS OF POPULATION GROWTH"
Reprinted here as it was released by the White House.

TO THE CONGRESS of the United States

In 1830 there were one billion people on the planet earth. By 1930 there were two billion, and by 1960 there were three billion. Today the world population is three and one half billion persons.

These statistics illustrate the dramatically increasing rate of population growth. It took many thousands of years to produce the first billion people; the next billion took a century; the third came after thirty years; the fourth will be produced in just fifteen.

If this rate of population growth continues, it is likely that the earth will contain over seven billion human beings by the end of this century. Over the next thirty years, in other words, the world's population could double. And a the end of that time, each new addition of one billion persons would not come over the millennia nor over a century nor even over a decade. If present trends were to continue until the year 2000, the eighth billion would be added in only five years and each additional billion in an even shorter period.

While there are a variety of opinions as to precisely how fast population will grow in the coming decades, most informed observers have a similar response to all such projections. They agree that population growth is among this most important issues we face. They agree that it can be met only if there is a great deal of advance planning. And they agree that the time for such planning is growing very short. It is for all these reasons that I address myself to the population problem in this message, first to its international dimensions and then to its domestic implications.

Continue reading Richard Nixon's Policies on International and Domestic Popultaion Control?

If no, drop.

If yes, go to www.boydgraves.com

Exhibit 8

RECEIVED
JAN 2 9 2001
LEONARD GREEN, Clerk

No. 99-4476

UNITED STATES COURT OF APPEALS
FOR THE SIXTH CIRCUIT

BOYD E. GRAVES,
Plaintiff-Appellant

V.

WILLIAM S. COHEN, et al,
Defendants-Appellees

PETITION FOR REHEARING EN BANC

The Appellant's exhibit (a DOD flowchart) requires a full review of the Court as to the issue of AIDS bioengineering. The appellant's exhibit alone passes the bar of frivolity identified by the lower court pursuant to Section 1915(e).

It is a travesty to the class of victims of this federal virus development program that this Court seeks to set aside the indisputable evidence of the true laboratory origin of the AIDS pandemic. The Special Virus program must be reviewed at some level of the trilateral government structure as compelled by the U. S. Constitution.

In any petition to the Supreme Court the class will seek to present the following question as representative of error of this Circuit:

Is the 1971 flowchart of the Special Virus program of the United States of America "sufficient evidence" to preclude a finding of frivolity pursuant to Section 1915(e) of the United States Code?

It is the class' position the flowchart and progress reports are conclusive proof of a federal program seeking to make a contagious cancer that selectively kills, by depleting the immune system. The Court en banc can not overlook the scientific significance of the completeness of the adjudication of the true origin of AIDS.

The complaint filed on September 28, 1998 requires a response as demanded by the United States Constitution consistent with Haines v. Kerner, **404 U.S. 519 - 521 (1972).** It is contrary to the social order of the common good of the United States to foster stealth programs that kill. The attached 5/15/00 letter from the National Institutes of Health confirms the reality of the program. The attached letter from the U.S. Attorney identifies the office of the true defendant best evidenced in P.L.91-213, signed March 16, 1970 by former President Richard M. Nixon. May God hath mercy on the United States, my country.

Respectfully submitted,

Boyd E. Graves, pro se, in forma pauperis
P.O. Box 332
Abilene, KS 67410
785-263-1871

Dated: January 22, 2001

CERTIFICATE OF SERVICE

I, Boyd E. Graves, do hereby certify that I served a copy of the foregoing PETITION FOR REHEARING ENBANC on Lisa Hammond Johnson, U.S. Attorney, sent via first class mail this ____day of January, 2001, postage prepaid.

BOYD E. GRAVES

TABLE II Analysis of Contracts by Activity

Phase I: SELECTION OF SPECIMENS AND DETECTION OF VIRUS OR VIRUS EXPRESSION
Step 1: SELECTION OF VIRUS SOURCES

Contractor	Cont. No.	Description of Work
Aichi Cancer Center	69-96	Seroepidemiological studies of BL, NPC in Southeast Asia
Baylor University	68-678	Seroepidemiology of cervical carcinoma
California SDPH	69-87	Human-feline retrospective epidemiological studies
California, Univ. of	72-2008	Serological studies on the relationship of HL-A type to cancer incidence
CDC	VCL-42	Epidemiological studies of selected leukemia cases (clusters)
Children's Hosp. (Phila)	66-477	Immunological studies of EBV-associated cancers
Georgetown University	65-53	Collaborative studies on populations at high risk to breast cancer
IARC	70-2076	Seroepidemiological study of BL and NPC in Southeast Asia and Africa
Inst. for Med. Res.	68-1000	Collaborative studies on populations at high risk to breast cancer
Jewish Hospital	72-2034	Genetic analysis of human cancer associated with chromosomal abnormalities
Johns Hopkins Univ.	71-2121	Seroepidemiology of cervical carcinoma
Karolinska Institute	69-2005	Immunological studies of EBV-associated cancers
Makerere University	67-47	Epidemiological studies of Burkitt Lymphoma in Uganda
Michigan Cancer Fdn.	71-2421	Collaborative studies on populations at high risk to breast cancer
Minnesota, Univ. of	71-2261	Immunological studies of high risk groups with immuno-deficiency diseases

TABLE II Analysis of Contracts by Activity

Phase I: SELECTION OF SPECIMENS AND DETECTION OF VIRUS OR VIRUS EXPRESSION
Step 1: SELECTION OF VIRUS SOURCES (continued)

Contractor	Cont. No.	Description of Work
Southern Calif., Univ. of	68-1030	Cancer surveillance and epidemiologic studies in Los Angeles County
Texas, University of	65-604	Serological studies of human leukemia, lymphoma, and solid tumors
Texas, University of	71-2135	Gather information on laboratory-acquired infections
Wolf R & D	71-2270	Develop computerized system for collection and storage of clinical and epidemiological information

TABLE II Analysis of Contracts by Activity

PHASE I: SELECTION OF SPECIMENS AND DETECTION OF VIRUS OR VIRUS EXPRESSION
Step 2: SOURCES OF VIRUS OR SUBVIRAL MATERIAL

Contractor	Cont. No.	Description of Work
Aichi Cancer Center	69-96	Acquire normal fetal and tumor tissue
Auerbach Associates	72-2023	Develop integrated systems for collection, storage, and distribution of resources.
Baylor University	68-678	Acquire clinical data and specimens of human neoplastic tissues
California, Univ. of	72-2008	Acquire clinical data and specimens of normal and neoplastic tissue
CDC	VCL-42	Collection of clinical data and specimens from human leukemias
Colorado University	69-2080	Pediatric and adult tumor specimens
Georgetown University	65-53	Acquire clinical data and human breast cancer specimens
Hospital for Sick Child.	65-97	Human leukemia and normal tissue collection
Howard University	70-2178	Acquire clinical data and human breast cancer specimens
IARC	70-2076	Acquire clinical data and specimens from BL and NPC.
Jewish Hospital	72-2034	Supply normal and neoplastic tissues from patients with various chromosomal abnormalities
Johns Hopkins Univ.	71-2109	Acquire clinical data and specimens from human leukemias and lymphomas
Karolinska Institute	69-2005	Acquire clinical data and specimens from BL, NPC, and leukemias
Makerere University	67-47	Collection of Burkitt Lymphoma specimens
Memorial Hospital (N.Y.)	71-2116	Acquire clinical data and specimens of human neoplastic tissue

TABLE II Analysis of Contracts by Activity

Phase I: SELECTION OF SPECIMENS AND DETECTION OF VIRUS OR VIRUS EXPRESSION
Step 2: SOURCES OF VIRUS OR SUBVIRAL MATERIAL

Contractor	Cont. No.	Description of Work
Memorial Hospital (N.Y.)	71-2194	Supply clinical data and specimens for human breast cancer studies
Michigan Cancer Fdn.	71-2421	Acquire clinical data and specimens for human breast cancer studies
Michigan, Univ. of	65-639	Collection of adult leukemia/lymphoma specimens
Minnesota, Univ. of	71-2261	Acquire clinical data and specimens from patients with immunological deficiency diseases
Montreal Children's Hosp.	65-1020	Collection of human leukemia and normal blood specimens and normal tissue
Padua, University of	68-1389	Collection of untreated human tumor and normal tissue specimens
Southern Calif., Univ. of	68-1030	Acquire clinical data and specimens of human neoplastic tissues
St. Joseph's Hospital	69-2074	Acquire clinical data and human sarcoma and breast cancer specimens
Texas, Univ. of	65-604	Acquire clinical data and specimens of human neoplastic tissue

TABLE II Analysis of Contracts by Activity

Phase 1: SELECTION OF SPECIMENS AND DETECTION OF VIRUS OR VIRUS EXPRESSION
Step 3: DETECTION OF VIRUS OR VIRUS EXPRESSION

Contractor	Cont. No.	Description of Work
Aichi Cancer Center	69-96	Detection using cell culture techniques-human tumors
Atomic Energy Commission	FS-7	Developmental research on immunological detection of tumor antigens--fetal antigens
Atomic Energy Commission	FS-13	Detection using immunological techniques
Baylor University	68-678	Detection using immunological and cell culture techniques
California, Univ. of (Davis)	70-2048	EM, biochemical, immunological techniques in comparative leukemia/sarcoma virus studies
California SDPH	68-997	Studies on the role of oncogenic viruses in cancer of man and domestic animals
California, Univ. of	72-2008	Detection using immunological techniques
California, Univ. of (also NBL, FS-8)	63-13	Tissue culture studies of normal and neoplastic human tissues
Children's Hospital (Phila)	66-477	Immunological detection of EBV-associated antigens in human cancer
Columbia University	70-2049	Screening human leukemia/lymphoma specimens with biochemical techniques
Cornell University	71-2508	Isolation, characterization of cat leukemia viruses
Cornell University	70-2224	Service - feline virus diagnostic laboratory
Einstein Medical College	65-612	Genetic studies on tumor/virus susceptibility
Flow Laboratory	71-2097	Immunological studies of mammalian Type C viruses
Georgetown University	65-53	Human breast cancer detection - biochemical techniques

TABLE II Analysis of Contracts by Activity

Phase I: SELECTION OF SPECIMENS AND DETECTION OF VIRUS OR VIRUS EXPRESSION
Step 3: DETECTION OF VIRUS OR VIRUS EXPRESSION (continued)

Contractor	Cont. No.	Description of Work
Harvard University	72-3246	Virus detection in non-human primates
Hazleton Labs	69-2079	Immunological/biochemical detection of virus in animal/human tumors
Howard University	70-2178	Immunological studies of human breast cancer
Huntingdon Research Ctr.	69-54	Immunological reagent production
IARC	70-2076	Immunological studies of BL, NPC
Indiana University	69-2048	Immunological characterization of avian RE tumor virus
Institute for Med. Res.	68-1000	Screening human and animal breast cancer specimens by EM, immunological techniques
Jackson Labs	67-744	Genetics of susceptibility to cancer in mice
Johns Hopkins University	71-2121	Immunological studies on Herpesvirus antigens in cervical carcinoma
Johns Hopkins University	71-2109	Immunological studies of human leukemia and lymphoma
Karolinska Institute	69-2005	Immunological studies of EBV-associated human neoplasia
Litton-Bionetics	71-2025	Screening of human/primate neoplastic tissue with biochemical techniques
Litton-Bionetics	69-2160	Detection using immunological techniques
Mason Research Institute	70-2204	Development of primate test systems for breast cancer virus detection
Meloy Labs	72-3202	Immunological studies of murine mammary tumor virus

TABLE II Analysis of Contracts by Activity

Phase I: SELECTION OF SPECIMENS AND DETECTION OF VIRUS OR VIRUS EXPRESSION
Step 3: DETECTION OF VIRUS OR VIRUS EXPRESSION (continued)

Contractor	Cont. No.	Description of Work
Meloy Labs	72-2006	Detection using immunological, biochemical and tissue culture techniques
Michigan Cancer Fdn.	71-2421	Virus detection in human breast cancer by biochemical techniques
Microbiological Assoc.	70-2068	Detection using immunological and cell culture techniques
Microbiological Assoc.	67-697	Bioassay of murine leukemia/sarcoma viruses
Microbiological Assoc.	67-700	Service - murine viral diagnostic reagents and testing
Minnesota, Univ. of	69-2061	Development of immunological tests for tumor antigens and antibodies
Minnesota, Univ. of	71-2261	Immunological and virological studies of immunodeficiency diseases
Netherlands Cancer Inst.	72-3260	Immunological detection of natural MTV expression
Ohio State University	65-1001	Immunological testing (PRILAT) for human tumor antigens and antibodies
Ohio State University	69-2233	See OSU 65-1001
Pennsylvania, Univ. of	65-1013	Studies of viruses associated with bovine leukemia
Pennsylvania State	70-2024	Biochemical, genetic studies of Herpes-type viruses
Pfizer, Chas., and Co.	67-1176	Detection of virus by EM-human and animal breast cancer
Princeton University	71-2372	Detection of cell membrane antigens by agglutination techniques
Public Health Res. Inst.	71-2129	Development of methods for isolation of virus from human neoplasia

TABLE II Analysis of Contracts by Activity

Phase I: SELECTION OF SPECIMENS AND DETECTION OF VIRUS OR VIRUS EXPRESSION
Step 3: DETECTION OF VIRUS OR VIRUS EXPRESSION (continued)

Contractor	Cont. No.	Description of Work
Robert Brigham Hosp.	71-2172	Detection using immunological methods
Rush-Presbyterian	71-2032	Immunological, biological, tissue culture studies of tumor viruses in non-human primates
Salk Institute	67-1147	Genetic, biochemical studies of viral-induced transformation
Salk Institute	72-3207	Immunological detection of tumor antigens
Scripps Clinic	72-3264	Development of new immunological technologies for detection of virus expression
Southern Calif., Univ. of	68-1030	Immunological studies of human fetal and tumor tissues
Stanford University	69-2053	Development of cell culture methods for human tissue
St. Joseph's Hospital	69-2074	Screening of human sarcomas by EM
St. Louis University	67-692	Detection using tissue culture and biochemical techniques
Tel Aviv University	72-3237	Biochemical detection of tumor viruses in human breast cancer
Texas, Univ. of	65-604	EM, tissue culture and immunological studies of human neoplastic tissues
Texas, Univ. of	71-2178	Immunological methods for detection of human tumor antigens and antibodies
Washington, Univ. of	71-2171	Development of immunological tests for tumor antigens and antibodies
Weizmann Institute	69-2014	Detection of tumor cell surface antigens by plant agglutinins

TABLE II Analysis of Contracts by Activity

Phase I: SELECTION OF SPECIMENS AND DETECTION OF VIRUS OR VIRUS EXPRESSION
Step 3: DETECTION OF VIRUS OR VIRUS EXPRESSION (continued)

Contractor	Cont. No.	Description of Work
Wisconsin, Univ. of	72-2022	Techniques for isolation and characterization of viral-induced tumor antigens
Wistar Institute	71-2092	Techniques for isolation and characterization of viral-induced tumor antigens

TABLE II Analysis of Contracts by Activity

Phase I: SELECTION OF SPECIMENS AND DETECTION OF VIRUS OR VIRUS EXPRESSION
Step 3: DETECTION OF VIRUS OR VIRUS EXPRESSION (continued)

Contractor	Cont. No.	Description of Work
Wisconsin, Univ. of	72-2022	Techniques for isolation and characterization of viral-induced tumor antigens
Wistar Institute	71-2092	Techniques for isolation and characterization of viral-induced tumor antigens

TABLE II Analysis of Contracts by Activity

Phase II-A: ESTABLISHMENT OF REPLICATION AND INITIAL CHARACTERIZATION
Step 1: ESTABLISH REPLICATION OF VIRUSES

Contractor	Cont. No.	Description of Work
Baylor University	68-678	Human leukemia transmission studies in non-human primates
Biolabs, Inc.	72-2068	Development and improvement of in vitro production of Herpes-type viruses
California SDPH	68-997	Isolation and characterization of feline tumor viruses
California, Univ. of	63-13	Development and evaluation of cell cultures for viral oncology research
California, Univ. of	70-2048	In vitro and in vivo studies of simian and feline virus infectivity and replication
California, Univ. of	72-2080	Development of methods for in vitro propagation of MTV
Children's Hosp. (Phila)	66-477	Developmental research on growth and replication of EBV in human cell lines
Cornell University	71-2508	Isolation and characterization of feline tumor viruses
Duke University	71-2132	Production and characterization of avian leukosis viruses
Electronucleonics	71-2253	Production and characterization of selected mammalian oncogenic viruses
Emory University	72-2301	Production of Herpes-type viruses
Flow Labs, Inc.	71-2097	Large-scale production of RNA tumor viruses for production of highly-specific diagnostic reagents
Harvard University	72-3246	Determination of host-range of primate oncogenic viruses
Hazleton Labs	69-2079	Isolation and production of ts-mutants of RNA tumor viruses
IARC	70-2076	Growth and replication of Herpes-type virus in nasopharyngeal carcinoma and Burkitt's lymphoma specimens

TABLE II Analysis of Contracts by Activity

Phase II-A: ESTABLISHMENT OF REPLICATION AND INITIAL CHARACTERIZATION
Step 1: ESTABLISH REPLICATION OF VIRUSES (continued)

Contractor	Cont. No.	Description of Work
Karolinska Institute	69-2005	Growth and replication of Herpes-type viruses
Life Sciences	69-63	Production and infectivity studies of Marek's disease virus
Litton-Bionetics	71-2025	Inoculation of nonhuman primates with various animal virus and human materials
Mason Research Inst.	70-2204	Studies on infectivity and effect of hormones on monkey mammary tumor virus replication
Medical Coll. of Wisconsin	68-1010	Stimulation of Type C virus production in human breast cancer cell line by various hormones
Meloy Labs	72-3202	In vitro and in vivo production of murine mammary tumor virus
Meloy Labs	72-2006	Growth and replication of mammalian Type C and syncytial viruses for tissue culture, biochemical studies of mammalian and avian tumor viruses
Miami University	70-2211	In vitro production of rat mammary tumor derived virus
Microbiological Assoc.	67-697	Cell-free transmission of murine mammary tumors with extracts from spontaneous tumors
Microbiological Assoc.	70-2068	Isolation, characterization, and transmission studies of mammalian and avian tumor viruses
Naval Biological Lab	FS-57	Studies of environmental factors influencing virus-host interactions--research on laboratory biohazards
North Dakota Univ.	66-8	Role of vectors in transmission, host range of tumor viruses
Ohio State Univ.	65-1001	Studies on factors affecting horizontal transmission of tumor viruses
Pennsylvania State	70-2024	Herpes-type virus replication in human cells

91

TABLE II Analysis of Contracts by Activity

Phase II-A: ESTABLISHMENT OF REPLICATION AND INITIAL CHARACTERIZATION
Step 1: ESTABLISH REPLICATION OF VIRUSES (continued)

Contractor	Cont. No.	Description of Work
Pennsylvania, Univ. of	65-1013	Experimental and natural transmission of bovine leukemia
Pfizer, Chas. & Co.	70-2080	Tissue culture production of Type C and Herpes-type viruses
Pfizer, Chas. & Co.	67-1176	Production of monkey and rat suspected oncogenic viruses
Rush-Presbyterian	71-2032	Mammalian tumor virus infectivity in nonhuman primates
Southern Calif., Univ. of	68-1030	Production of mammalian RNA tumor virus and candidate human agents
Southwest Foundation	71-2348	Study of latent virus infection and transmission--research on laboratory biohazards
St. Jude's Hospital	71-2134	Isolation and characterization of oncogenic Herpes-type viruses
St. Louis University	67-692	In vitro cultivation of various mammalian Type C tumor viruses for biochemical studies
Texas, University of	65-604	Infectivity, oncogenicity and host range studies of hamster sarcoma virus; isolation and characterization of candidate human Type C oncogenic virus
University Labs	66-1133	In vitro and in vivo production of murine and avian tumor viruses

92

TABLE II Analysis of Contracts by Activity

Phase II-A: ESTABLISHMENT OF REPLICATION AND INITIAL CHARACTERIZATION
Step 2: INITIAL CHARACTERIZATION

Contractor	Cont. No.	Description of Work
Baylor University	68-678	Comparative characterization of human Herpes-type viruses
California, Univ. (Davis)	70-2048	Comparative studies on simian leukemia/sarcoma viruses
Children's Hospital (Phila)	66-477	Immunological, tissue culture characterization of EBV
Columbia University	70-2049	Biochemical characterization of mammalian Type C viruses
Cornell University	71-2508	Immunological characterization of feline tumor virus isolates
Duke University	71-2132	Immunological characterization of RNA tumor viruses
Emory University	72-2301	Serological studies on Herpes-type virus antigens
Flow Labs	71-2097	Immunological, tissue culture studies of mammalian DNA and RNA oncogenic viruses.
Harvard University	72-3246	Characterization of simian oncogenic Herpes-type viruses
Hazleton Labs	69-2079	Characterization of ts mutants of RNA tumor viruses
IARC	70-2076	Isolation and characterization of Herpes-type virus in cultures of Burkitt's lymphoma and nasopharyngeal carcinoma
Indiana University	69-2048	Immunological, biochemical characterization of avian RE virus
Karolinska Institute	69-2005	Immunological and biochemical characterization of EBV
Life Sciences	69-63	*In vitro* and *in vivo* studies of Marek's disease virus
Meloy Labs	72-2006	Biochemical characterization of mammalian Type C viruses

TABLE II Analysis of Contracts by Activity

Phase II-A: ESTABLISHMENT OF REPLICATION AND INITIAL CHARACTERIZATION
Step 2: INITIAL CHARACTERIZATION (continued)

Contractor	Cont. No.	Description of Work
Meloy Labs	72-3202	Immunological characterization of murine MTV
Miami University	70-2211	EM, biological, biochemical characterization of rat mammary tumor derived virus
Microbiological Assoc.	70-2068	Immunological characterization of mammalian Type C tumor viruses
Pennsylvania State	70-2024	Genetic, biochemical studies of cells "transformed" by viral (Herpes-simplex)--chemical cocarcinogenesis
Pennsylvania, Univ. of	65-1013	Characterization of Type C virus associated with bovine leukemia
Rush-Presbyterian	71-2032	Immunological, biological characterization of Herpes viruses of nonhuman primates
Scripps Clinic & Res. Fdn.	72-3264	Development and improvement of specific viral diagnostic reagents
Southern Cal., Univ. of	68-1030	Immunological characterization of mammalian tumor viruses
St. Jude Children's Res. Hosp.	71-2134	Characterization of suspected oncogenic Herpes-type viruses
St. Louis University	67-692	Biochemical characterization of oncogenic RNA and DNA viruses
Texas, Univ. of	65-604	Studies on the relationship of animal tumor viruses to human leukemia and lymphomas; characterization of candidate human Type C oncogenic virus
TRW	70-2200	Improvement of methods for production of specific viral diagnostic reagents
Wisconsin, Univ. of	72-2022	Isolation and characterization of subunits of RNA tumor viruses
Wistar Institute	71-2092	Isolation and characterization of oncogenic DNA and RNA virus-induced tumor antigens

TABLE II Analysis of Contracts by Activity

Phase II-B: REPLICATION AND CHARACTERIZATION OF VIRUS EXPRESSION
Step 1: INDUCE VIRAL REPLICATION OF WHOLE VIRUS OR TRANSMISSION OF EXPRESSION

Contractor	Cont. No.	Description of Work
Aichi Cancer Center	69-96	Tissue culture methods to induce virus replication in human tumor cells
AEC	FS-13	Induction of virus expression by cell fusion techniques
Baylor University	72-2058	Development of "nonsense" suppressor mutant cell lines for viral genome characterization
California SDPH	68-997	Establish cultures from tumors of domestic animals and attempt to rescue defective viral genome
California, Univ. of	63-13	Tissue culture of human neoplastic tissue for induction of virus replication
Flow Labs	71-2097	Cell hybridization techniques to rescue "defective" viruses
Hazleton Laboratories	69-2079	Transmission of oncogenic virus expression in selected cell systems
Hôpital St. Louis	72-3263	Induction and transmission of oncogenic virus expression in human cells
Illinois, Univ. of	72-2031	Development of methods for recognition of virus expression
Medical Coll. of Wisc.	68-1010	Co-cultivation of human breast cancer and hormone-secreting cell lines
Meloy Labs	72-2006	Characterization of non-producer transformed cells; virus rescue techniques
Microbiological Assoc.	67-697	Effect of hormones on virus expression
Microbiological Assoc.	70-2068	Studies on Type C viral genome expression--effect of chemical carcinogens
Pennsylvania State	70-2024	Induction and maintenance of human Herpes-type virus oncogenicity
Public Health Res.	71-2129	Rescue of viruses from human tumors

95

TABLE II Analysis of Contracts by Activity

Phase II-B: REPLICATION AND CHARACTERIZATION OF VIRUS EXPRESSION
Step 1: INDUCE VIRAL REPLICATION OF WHOLE VIRUS OR TRANSMISSION OF EXPRESSION (continued)

Contractor	Cont. No.	Description of Work
Salk Institute	67-1147	Studies on the activation of Type C virus genome by polyoma virus
Southern Cal., Univ. of	68-1030	Studies of virus expression in human fetal and tumor tissue
Southern Cal., Univ. of	72-2032	Methods for recognition and/or rescue of tumor virus expression
Stanford University	69-2053	Tissue culture, biochemical methods to induce virus replication in human tumor cells
Tel Aviv University	72-3237	Development of methods for recognition of virus expression in human breast cancer
Texas, Univ. of	65-604	Attempts to induce viral replication in human cell lines

96

TABLE II Analysis of Contracts by Activity

Phase II-B: REPLICATION AND CHARACTERIZATION OF VIRUS EXPRESSION
Step 2: INITIAL CHARACTERIZATION

Contractor	Cont. No.	Description of Work
Atomic Energy Commission	FS-13	Biochemical studies on regulation of gene expression
California, Univ. of	71-2147	Molecular studies of avian tumor-virus-associated enzymes
California, Univ. of	71-2173	Molecular studies of the structure of oncogenic viruses and characterization of viral-specific enzymes
California, Univ. of	72-3236	Characterization of growth regulatory mechanism in normal and neoplastic cells
Columbia University	70-2049	Biochemical characterization of viral-specific enzymes
Einstein College of Med.	71-2251	Biochemical characterization of viral-specific enzymes and other proteins
Flow Labs	71-2097	Characterization of tumor virus expression in mammalian systems
Hazleton Laboratories	69-2079	Characterization of cellular and subcellular alterations in viral transformation
Massachusetts Gen. Hos.	71-2174	Characterization of nucleic acids and proteins of AMV
Mass. Inst. of Technol.	71-2149	Biochemical characterization of viral-specific enzymes
Meloy Labs	72-2006	Multidisciplinary approaches to characterization of virus-expression and mediators of replication
Microbiological Assoc.	70-2068	Immunological identification of antigens related to known tumor viruses
North Carolina, Univ. of	72-3228	Biochemical identification of DNA viral genome in human cells
Oregon State Univ.	71-2175	Correlation of ultrastructural and biochemical changes associated with transformation by viruses
Princeton Univ.	71-2372	Characterization of cell membrane changes in malignant transformation

TABLE II Analysis of Contracts by Activity

Phase II-B: REPLICATION AND CHARACTERIZATION OF VIRUS EXPRESSION
Step 2: INITIAL CHARACTERIZATION (continued)

Contractor	Cont. No.	Description of Work
Public Health Res.	71-2129	Identification of cellular and subcellular alterations characteristic of malignant transformation
Public Health Res.	72-2028	Characterization of cell membrane changes in malignant transformation
Scripps Clinic & Res. Fdn.	72-3264	Development and improvement of immunochemical methods for the detection of cell membrane changes induced by oncogenic viruses
Southern Calif., Univ. of	68-1030	Immunological identification of antigens related to known tumor viruses
Southern Calif., Univ. of	72-2032	Establishment of methods for identification of virus-induced transformation
St. Louis Univ.	67-692	Multidisciplinary approaches to characterization of oncogenic virus expression and mediators of replication
Tel Aviv Univ.	72-3237	Biochemical identification of subviral expression in breast cancer
Texas, Univ. of	65-604	Characterization of <u>in vitro</u> transformation of human sarcoma cells
Weizmann Institute	69-2014	Improvement of immunochemical methods for detection of cell membrane changes in viral transformation
Wisconsin, Univ. of	72-2022	Development of immunochemical reagents and tests for detection of virus expression in chemically-induced tumors

TABLE II Analysis of Contracts by Activity

Phase III-A: COMPLETE CHARACTERIZATION AND DEFINITION OF PRESUMPTIVE DISEASE RELATIONSHIPS
Step 1: PRESUMPTIVE DISEASE RELATIONSHIPS

Contractor	Cont. No.	Description of Work
Atomic Energy Comm.	FS-13	Interaction of RNA tumor viruses and host immune mechanism; studies on relationship of embryogenesis and carcinogenesis
Baylor University	68-678	Studies on presumptive disease relationships of MTV
California SDPH	68-997	Serological testing of host reactions to tumor virus antigens
Children's Hosp. (D.C.)	72-2071	Cell-mediated immunity to human cancers
Children's Hosp. (Phila.)	66-477	Relationship of EBV to human lymphoma
Columbia University	70-2049	Biochemical studies on relationship of Type C and Type B viruses to human leukemia/sarcoma and breast cancer
Einstein Medical College	65-612	Genetic studies on tumor/virus susceptibility
Emory University	72-2301	Determination of host response to Herpes-type virus in cervical cancer
Flow Laboratories	71-2097	Complete characterization of RNA and DNA viruses and viral antigens in mammalian tumors
Georgetown University	65-53	Studies on Type B particles associated with human breast cancer
George Washington Univ.	72-3251	Cell-mediated immunity to human cancers
Howard University	70-2178	Cell-mediated immunity to human cancers
IARC	70-2076	Seroepidemiological studies of Burkitt's lymphoma, NPC
Institute for Med. Res.	68-1000	Studies on Type B particles associated with human breast cancer
Jackson Labs	67-744	Natural occurrence of RNA tumor viruses and host gene control of virus expression

TABLE II Analysis of Contracts by Activity

Phase III-A: COMPLETE CHARACTERIZATION AND DEFINITION OF PRESUMPTIVE DISEASE RELATIONSHIPS (continued)
Step 1: PRESUMPTIVE DISEASE RELATIONSHIP

Contractor	Cont. No.	Description of Work
Jewish Hospital	72-2034	Relationship of chromosomal abnormalities to susceptibility to cancer and viral transformation
Johns Hopkins Univ.	71-2121	Studies on the relationships of Herpes simplex type II to cervical carcinoma
Karolinska Institute	69-2005	Immunological studies on the etiology of EBV-associated diseases
Life Sciences	69-63	Studies on Marek's disease Herpes virus
Litton-Bionetics	71-2025	Biochemical studies on relationship of Type C viruses to human leukemia
iiversity	67-47	Epidemiological studies on role of EBV in Burkitt's lymphoma
Hospital	72-2012	Interaction of oncogenic viruses and host immune mechanisms; relationship of immunological competence and viral carcinogenesis
oratories	72-2006	Biochemical studies on relationship of Type C viruses to human leukemia and sarcoma
iversity	67-1187	Immunological responses in avian tumor virus infection
ological Assoc.	70-2068	Evaluation of cocarcinogenic factors in viral oncogenesis
ological Assoc.	67-697	Type C virus antigen expression during embryogenesis and in spontaneous cancers
ota, Univ. of	69-2061	Immunologic evaluation of host response to human tumors
, Univ. of	71-2056	Isolation and characterization of Herpes simplex virus-induced antigens
rlands Cancer Inst.	72-3260	Determination of natural route of infection of MTV

TABLE II Analysis of Contracts by Activity

Phase III-A: COMPLETE CHARACTERIZATION AND DEFINITION OF PRESUMPTIVE DISEASE RELATIONSHIPS (continued)
Step 1: PRESUMPTIVE DISEASE RELATIONSHIP

Contractor	Cont. No.	Description of Work
New York Med. College	72-3289	Immunopathology of human breast cancer
Ohio State University	65-1001	Determination of immune response to viral antigens in model systems
Pennsylvania State	70-2024	Effect of cocarcinogens on oncogenic potential of human viruses
Robert B. Brigham	71-2172	Immunologic evaluation of host response to viral antigens in model systems
Rutgers, The State Univ.	71-2077	Relationship of presumed non-oncogenic agents to cancer induction
Salk Institute	72-3207	Immunologic studies on host reaction to viral antigens
Southern Calif., Univ. of	68-1030	Possible role of animal tumor viruses, environmental cocarcinogens
Texas, Univ. of	71-2178	Immunologic studies on host reaction to tumor antigens
Texas, Univ. of	72-3262	Determination of host reaction to murine leukemia virus antigens in human cancer patients
Washington, Univ. of	71-2171	Immunologic reactivity to tumor antigens in patients with various malignancies
Washington, Univ. of	72-2037	Immunologic reactivity to canine sarcomas
Wisconsin, Univ. of	72-2022	Development and improvement of methods for the detection and quantitation of immunity to oncogenic viruses and viral antigens

101

TABLE II Analysis of Contracts by Activities

PHASE III-A: COMPLETE CHARACTERIZATION AND DEFINITION OF PRESUMPTIVE DISEASE RELATIONSHIPS
Step 2: COMPLETE CHARACTERIZATION

Contractor	Cont. No.	Description of Work
Columbia Univ.	70-2049	Biochemical characterization of oncogenic viruses
Flow Laboratories	71-2097	Biochemical, biophysical and immunologic characterization of oncogenic viruses
Karolinska Institute	69-2005	Immunological characterization of EBV
Life Sciences	69-63	Biological characterization of MDHV
Meloy Laboratories	72-2006	Biochemical, biophysical and immunologic characterization of oncogenic viruses

102

TABLE II Analysis of Contracts by Activity

Phase III-B: COMPLETE CHARACTERIZATION: DEMONSTRATION OF VIRUS-MEDIATED FUNCTIONS ESSENTIAL FOR INDUCTION AND MAINTENANCE OF NEOPLASIA

Contractor	Cont. No.	Description of Work
California, Univ. of	71-2147	Biochemical determination of viral gene expression (molecular hybridization)
California, Univ. of	72-3226	Molecular hybridization studies of human cancers
Columbia University	70-2049	Search for specific viral gene expressions in human cancer
Einstein Coll. of Med.	71-2251	Determine molecular pathways of oncogenic virus expression
Life Sciences, Inc.	69-63	Co-carcinogenic factors in the etiology of Marek's disease
Mass. Gen. Hospital	71-2174	Complete characterization of oncogenic viral nucleic acids
Mass. Inst. of Tech.	71-2149	Determine the nature of oncogenic viral gene expression
Meloy Laboratories	72-2006	Molecular hybridization studies on human cancers
North Carolina, Univ. of	72-3228	Molecular hybridization studies of human leukemia and lymphoma
Pub. Health Res. Inst.	71-2129	Characterization of the specific membrane changes associated with oncogenic virus transformation
St. Louis University	67-692	Search for specific viral gene expressions in human cancer

TABLE II Analysis of Contracts by Activity

Phase IV-A: IMMUNOLOGICAL CONTROL
Step 1: DETERMINE SUITABLE IMMUNOLOGICAL CONTROL

Contractor	Cont. No.	Description of Work
Health Res. Inc.	72-2014	Evaluation of neuraminidase-treatment to enhance tumor cell immunogenicity
Johns Hopkins Univ.	71-2109	Evaluation of methods for monitoring immune responses of cancer patients
Meloy Laboratories	72-2020	Evaluation of various approaches to immunotherapy in model systems
Microbiological Assoc.	70-2068	Evaluation of viral vaccines and interferons in the protection against chemically-induced neoplasms
Merck and Co.	71-2059	Developmental research for virus vaccine production
Res. Fdn. of State of New York	71-2137	Clinical studies on enhancement of tumor immunity
Texas, Univ. of	72-3260	Evaluation of viral vaccines in the treatment of human leukemia/lymphoma

TABLE II Analysis of Contracts by Activity

Phase IV-B: BIOCHEMICAL CONTROL
Step 1: DETERMINE SUITABLE METHODS FOR BIOCHEMICAL CONTROL

Contractor	Cont. No.	Description of Work
St. Louis University	67-692	Screening of various chemicals as inhibitors of polymerases

Exhibit 10

NATIONAL SECURITY COUNCIL MEMORANDUM-46

(SECRET) MARCH 17, 1978

Interdepartmental Review Memorandum NSCM-46
TO: The Secretary of State
The Secretary of Defense
The Director of Central Intelligence

SUBJECT: Black Africa and the U.S. Black Movement

The President has directed that a comprehensive review be made of current developments in Black Africa from the point of view of their possible impacts on the black movement in the United States. The review should consider:

1. Long-term tendencies of social and political developments and the degree to which they are consistent with or contradict the U.S. interests.
2. Proposals for durable contacts between radical African leaders and leftist leaders of the U.S. black community.
3. Appropriate steps to be taken inside and outside the country in order to inhibit any pressure by radical African leaders and organizations on the U.S. black community for the latter to exert influence on the policy of the Administration toward Africa.

The President has directed that the NSC Interdepartmental Group for Africa perform this review. The review should be forwarded to the NSC Political Analysis Committee by April 20.

(signed)

Zbigniew Brezinski

cc: The Secretary of the Treasury
The Secretary of Commerce
The Attorney General
The Chairman Joint Chiefs of Staff

To the people of this world:

Sufficient evidence exists to make a reasonable determination that World powers believed it was important to depopulate Africa. If that were true, there would exist a contingency plan for the backlash of negative African American sentiment once the truth were known. We believe we can show the AIDS virus is a prototype virus of the "Special Virus" program of the United States. We believe the 1971 AIDS Flow Chart adequately expresses the logic of the United States to create an immune suppressing virus that has an affinity toward Africans and their descendants. We also believe the fifteen (15) progress reports of the "Special Virus" accurately portray the data, experiments, contracts and contractors of the World's hunt for a predatory virus.

SECRET: NATIONAL SECURITY COUNCIL INTERDEPARTMENTAL GROUP FOR AFRICA

STUDY RESPONSE TO PRESIDENTIAL SECURITY REVIEW MEMORANDUM 1 NSCM-46

BLACK AFRICA AND THE U.S. BLACK MOVEMENT

I. (most text blacked out) . . .and whose importance for the United States is on the increase.

II. A. U.S. INTERESTS IN BLACK AFRICA

A multiplicity of interests influences the U.S. attitude toward black Africa. The most important of these interests can be summarized as follows:

1. POLITICAL

If black African states assume attitudes hostile to the U.S. national interest, our policy toward the white regimes; which is a key element in our relations with the black states, may be subjected by the latter to great pressure for fundamental change. Thus the West may face a real danger of being deprived of access to the enormous raw material resources of southern Africa which are viral for our defense needs as well as losing control over the Cape sea routes by which approximately 65% of Middle Eastern oil is supplied to Western Europe.

Moreover, such a development may bring about internal political difficulties by intensifying the activity of the black movement in the United States itself.

It should also be borne in mind that black Africa is an integral part of a continent where tribal and regional discord, economic backwardness, inadequate infrastructures, drought, and famine, are constant features of the scene. In conjunction with the artificial borders imposed by the former colonial powers, guerilla warfare in Rhodesia and widespread indignation against apartheid in South Africa, the above factors provide the communist states with ample opportunities for furthering their aims. This must necessarily redound to the detriment of U.S. political interests.

2. ECONOMIC

Black Africa is increasingly becoming an outlet for U.S. exports and investment. The mineral resources of the area continue to be of great value for the normal functioning of industry in the United States and allied countries. In 1977, U.S. direct investment in black Africa totaled about $1.8 billion and exports $2.2 billion. New prospect of substantial profits would continue to develop in the countries concerned.

IV. BLACK AFRICA AND THE U.S. BLACK MOVEMENT

Apart from the above-mentioned factors adverse to U.S. strategic interests, the nationalist liberation movement in black Africa can act as a catalyst with far reaching effects on the American black community by stimulating its organizational consolidation and by inducing radical actions. Such a result would be likely as Zaire went the way of Angola and Mozambique. An occurrence of the events of 1967-68 would do grievous harm to U.S. prestige, especially in view of the concern of the present Administration with human rights issues. Moreover, the Administration would have to take specific steps to stabilize the situation. Such steps might be misunderstood both inside and outside the United States.

In order to prevent such a trend and protect U.S. national security interests, it would appear essential to (text missing) African Nationalist Movement. In elaborating U.S. policy toward black Africa, due weight must be given to the fact that there are 25 millions American blacks whose roots are African and who consciously or subconsciously sympathies with African nationalism.

The living conditions of the black population should also be taken into account. Immense advances in the field are accompanied by a long-lasting high rate of unemployment, especially among the youth and by poverty and dissatisfaction with government social welfare standards.
These factors taken together may provide a basis for joint actions of a concrete nature by the African nationalist movement and the U.S. black community. Basically, actions would take the form of demonstrations and public protests, but the likelihood of violence cannot be excluded. There would also be attempts to coordinate their political activity both locally and in international organizations.

Inside the United States these actions could include protest demonstrations against our policy toward South Africa accompanied by demand for boycotting corporations and banks which maintain links with that country; attempts to establish a permanent black lobby in Congress including activist leftist radical groups and black legislators; the reemergence of Pan-African ideals; resumption of protest marches recalling the days of Martin Luther King; renewal of the extremist idea national idea of establishing an "African Republic" on American soil. Finally, leftist radical elements of the black community could resume extremist actions in the style of the defunct Black Panther Party.

Internationally, damage could be dome to the United States by coordinated activity of African states designed to condemn U.S. policy toward South Africa, and initiate discussions on the U.S. racial issue at the United Nations where the African representation constitutes a powerful bloc with about one third of all the votes.

A menace to U.S. economic interests, though not a critical one, could be posed by a boycott by Black African states against American companies which maintain contact with South Africa and Rhodesia. If the idea of economic assistance to black Americans shared by some African regimes could be realized by their placing orders in the United States mainly with companies owned by blacks, they could gain a limited influence on the U.S. black community.

In the above context, we must envisage the possibility, however remote, that black Americans interested in African affairs may refocus their attention on the Arab-Israeli conflict. Taking into account; the African descent of American blacks it is reasonable to anticipate that their sympathies would lie with the Arabs who are closer to them in spirit and in some case related to them by blood. Black involvement in lobbying to support the Arabs may lead to serious dissention between American black and Jews. The likelihood of extremist actions by either side is negligible, but . . . . (text cut off)
(text returns)

3. Political options

In the context of long-term strategy, the United States can not afford a radical change in the fundamentals of its African policy, which is designed for maximum protection of national security. In the present case, emphasis is laid on the importance of Black Africa for U.S. political, economic and military interests.

RECOMMENDATIONS

In weighing the range of U.S. interests in Black Africa, basic recommendations arranged without intent to imply priority are:

1. Specific steps should be taken with the help of appropriate government agencies to inhibit coordinated activity of the Black Movement in the United States.

2. Special clandestine operations should be launched by the CIA to generate mistrust and hostility in American and world opinion against joint activity of the two forces, and (foster?) division among Black African radical national groups and their leaders.

3. U.S. embassies to Black African countries specially interested in southern Africa must be highly circumspect in view of the activity of certain political circles and influential individuals opposing the objectives and methods of U.S. policy toward South Africa. It must be kept in mind that the failure of U.S. strategy in South Africa would adversely affect American standing throughout the world. In addition, this would mean a significant diminution of U.S. influence in Africa and the emergence of new difficulties in our internal situation due to worsening economic prospects.

4. The FBI should mount surveillance operations against Black African representatives and collect sensitive information on those, especially at the U.N., who oppose U.S. policy toward South Africa. The information should include facts on their links with the leaders of the Black movement in the United States, thus making possible at least partial neutralization of the adverse effects of their activity.

V. TRENDS IN THE AMERICAN BLACK MOVEMENT

In connection with our African policy, it is highly important to evaluate correctly the present state of the Black movement in the Untied States and basing ourselves on all available information, to try to devise a course for its future development. Such an approach is strongly suggested by our perception of the fact that American Blacks form a single ethnic group potentially capable of causing extreme instability in our strategy toward South Africa. This may lead to critical differences between the United States and Black Africa in particular. It would also encourage the Soviet Union to step up its interference in the region. Finally, it would pose a serious threat to the delicate structure of race relations within the United States. All the above considerations give rise to concern for the future security of the United States.

(text returns)

undergone considerable changes. The principle changes are as follows:

(remaining text lost to the right margin)

-Social and economic issues have supplanted political aims as the main preoccupations of the movement. ( ) actions formerly planned on a nationwide scale are now being organized locally.

-Fragmentation and a lack of organizational unity within movement.

-Sharp social stratification of the Black population and lack of policy options which could reunite them.

-Want of a national leader of standing comparable to Martin Luther King.

B. THE RANGE OF POLICY OPTIONS

The concern for the future security of the United States makes necessary the range of policy options. Arranged without intent imply priority they are:

(a) to enlarge programs, within the framework of the present budget, for the improvement of the social and economic welfare of American Blacks in order to ensure continuing development of present trends in the Black movement;

(b) to elaborate and bring into effect a special program designed to perpetuate division in the Black movement and neutralize the most active groups of leftist radical organizations representing different social strata of the Black community: to encourage division in Black circles;

(c) to preserve the present climate which inhibits the emergence from within the Black leadership of a person capable of exerting nationwide appeal;

(d) to work out and realize preventive operations in order to impede durable ties between U.S Black organizations and radical groups in African states;

(e) to support actions designed to sharpen social stratification in the Black community which would lead to the widening and perpetuation of the gap between successful educated Blacks and the poor, giving rise to growing antagonism between different Black groups and a weakening of the movement as a whole;

(f) to facilitate the greatest possible expansion of Black business by granting government contracts and loans with favorable terms to Black businessmen;

(g) to take every possible means through the AFL-CIO leaders to counteract the increasing influence of Black labor organizations which function in all major unions and in particular, the National Coalition of Black Trade Union and its leadership including the creation of real preference for adverse and hostile reaction among White trade unionists to demands for improvement of social and economic welfare of the Blacks;

(h) to support the nomination at federal and local levels of loyal Black public figures to elective offices, to government agencies and the Court. This would promote the achievement of a twofold purpose...........

PHASE II

CORRESPONDENCE

www.boydgraves.com

"By what natural mechanism does an icelandic sheep disease VISNA appear in the genetic sequence of HIV?"

Boyd E. Graves, J.D.
One of the questions presented to the United States Supreme Court April 11, 2001

Readers Note: The correspondences are reproduced with the principals' business contact information as originally written. In some cases this contact information has changed. Please verify addresses before sending official correspondences.

Post: Boyd E. Graves, J.D.
c/o Zygote Media
PO Box 332
Abilene, KS 67410
Email: boyded2001@yahoo.com
Phone: 800-257-9387

NATIONAL INSTITUTES OF HEALTH

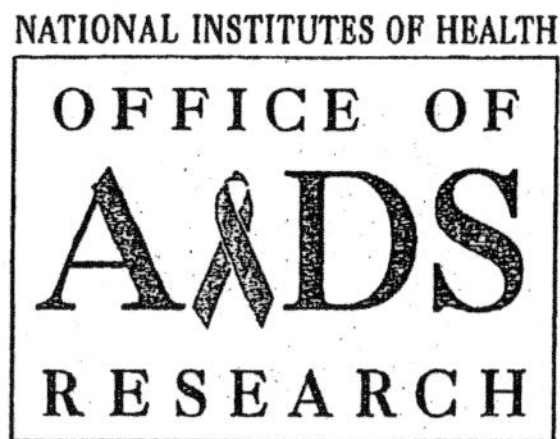

oria A. Cargill, M.D., M.S.C.E.
Medical Officer

Building 2, Room 4E20
Two Center Drive
MSC 0255
Bethesda, MD 20892
TEL: 301 402 2932
FAX: 301 496 4843
E-Mail: cargillv@od.nih.gov

MAY 15 2000

Boyd E. Graves, Esq.
2700 Washington Street
Apartment 1402
Cleveland, Ohio 44113

Dear Mr. Graves:

I am writing to follow up on our telephone conversation last week regarding the flow chart on the Human Immunodeficiency Virus (HIV). As I indicated on the telephone, I called you in response to a telephone contact you made to Ms. Linda Jackson in our office regarding this flow chart. Using the information you provided, I located and read the materials on your web sites:

http://www.alltheweb.com
http://www.aidsbiowar.com.

Subsequent to our telephone conversation, I located and contacted Mrs. Judith Grossberg, the librarian at the National Cancer Institute archives. She confirmed the presence of the reports on the Special Virus, and obtained report #8. I have reviewed this report, as well as the flow chart, to which you referred.

I have referred this matter, and the report citations to the National Cancer Institute, as the research was conducted within their Institute. I have provided for them the specific report numbers, and citations which you shared with me, as well as your web sites indicated above.

Thank you for your interest in AIDS.

Sincerely,

Victoria A. Cargill, MD. MSCE

Victoria A. Cargill, M.D., M.S.C.E.
Medical Officer

JAMES A. TRAFICANT, JR.
17TH DISTRICT OHIO

COMMITTEE
TRANSPORTATION AND
INFRASTRUCTURE
SUBCOMMITTEES
RANKING DEMOCRAT: PUBLIC BUILDINGS
AND ECONOMIC DEVELOPMENT
AVIATION

COMMITTEE:
SCIENCE
SUBCOMMITTEE:
SPACE

Congress of the United States
House of Representatives
Washington, DC 20515–3517

2446 RAYBURN HOUSE OFFICE BUILDING
WASHINGTON, DC 20515
(202) 225-5261

125 MARKET STREET
YOUNGSTOWN, OH 44503
(216) 743-1914

5555 YOUNGSTOWN-WARREN ROAD
SUITE 503
NILES, OH 44446
(216) 652-5649

109 WEST 3RD STREET
EAST LIVERPOOL, OH 43920
(216) 385-5921

May 23, 1997

Exhibit 2

The Honorable Dan Burton
Chairman
Committee on Government
Reform and Oversight
2157 Rayburn House Office Building
Washington, D.C. 20515

Dear Chairman Burton:

I am writing to you regarding the case of Boyd E. Graves, a constituent of mine who faced discrimination in his quest for a job at the United States Architectural and Transportation Barriers Compliance Board (Access Board).

He is a Naval Academy graduate and a lawyer. I have been working on Ed's discrimination case since October, 1995.

Mr. Graves applied for a job at the United States Architectural and Transportation Barriers Compliance Board (the Access Board) in March, 1995. He was denied the position despite the fact that he is disabled (HIV+) and applicants were eligible for non-competitive appointment, not requiring competitive status. In other words, he should have been given equal treatment for his disability. Mr. Graves provided all the necessary documentation certifying his disability. When Mr. Graves called to check on the status of his application, he was told he was not under consideration

The Access Board selected someone with less education and experience and no disability. The Access Board advertised two more positions available and Mr. Graves submitted applications and resumes. The EEO officer of the Access Board, told Mr. Graves in May of 1995 that the Access Board was aware he had applied for the second position and would not award him the position because he filed a complaint regarding the first non-hiring. She also said they would interview him so as not to face a second discrimination charge. He was interviewed for the second opening in June 23, 1995. On July 10, 1995, he was contacted by the Board by phone letting him know he was still under consideration for the second position. On July 30, 1995, the Board informed Mr. Graves that while he was qualified for the position, they had hired another candidate.

On August 3, 1995, Mr. Graves filed a discrimination

Exhibit 2

complaint with the EEO officer of the Access Board. He charged two separate complaints from two posted job openings. He charged that the Board eliminated him from any position upon learning of his disability. He also charges that the Board retaliated against him for exercising his civil rights by not considering his candidacy for the second position.

On December 8, 1995, Mr. Graves received notification from the Board acknowledging receipt of his complaint. It took nearly two months for GSA to forward the complaint to the Access Board on October 11. The Access Board arranged to have an outside entity investigate the complaint which would take 180 days. The letter informed Mr. Graves of all his options according to the law. Mr. Graves responded with a letter five days later stating that it took GSA six weeks to forward the complaint to the Board and another six weeks to notify Mr. Graves. In the letter, Mr. Graves speculates that his October letter to the President prompted an inquiry by the White House and the Board only responded because of that inquiry. Mr. Graves also questioned the validity and conviction of the investigation in light of the response time.

In response, the Board notified Ed that GSA forwarded the original which was stamped with a date of Sept. 18, 1995 on one page as a date of receipt and with a date of August 21, 1995 on another page. Mr. Graves claims to have hand delivered the letter to GSA on August 17, 1995. The Board also stated that the Office of Federal Operations in the Equal Employment Opportunity Commission would be handling the case for them.

On December 18, 1995, Mr. Graves spoke with James Raggio, General Counsel of the Board. During the conversation, Mr. Graves raised several points with Mr. Raggio. These include 1) Mr. Graves was uncomfortable with Raggio familiarity with the investigator, James Popiden. 2) Mr. Graves's concerns regarding the date of receipt discrepancies, and 3) an investigation into why the letter has two dates.

In January, our office received response from the Board to our letter. They stated, "(w)e received many applications from qualified individuals for the two positions. We followed merit system principles in selecting applicants for each position. The individuals who were selected were highly qualified. There was no unlawful discrimination against any of the applicants who were not selected.

Mr. Graves filed several FOIA requests for information regarding his case from the Board. Among these documents, he found evidence that: some files where shredded by the EEO officer, the list of potential hires show Mr. Graves was never under consideration for the second position, one applicant was hired despite an incomplete application, favoritism to candidates familiar with the Board, and other inaccuracies in the record.

On March 26, 1996, Mr. Graves and the Board were supposed to meet at the EEOC to discuss a resolution. However, on March 21, 1996, that meeting was cancelled because the Board did not want to discuss resolution. At this time, the EEOC stated they would not get involved anymore. Mr. Graves has filed an appeal to the EEOC's decision.

Mr. Graves's case goes on and on and the has involved several federal agencies and departments including the Office of Special Counsel, the Federal Bureau of Investigations, the Department of Justice, the Commission on Civil Rights and the Department of Education.

Most importantly to Mr. Graves is the EEOC's failure to include several items on the record in the Findings and Conclusions of the EEOC's issuance on December 17, 1996. In this regard, the United States Commission on Civil Rights recently referred Mr. Graves's case back to the EEOC for further review. A final decision is pending.

Ironically, the United States Architectural and Transportation Barriers Compliance Board was chartered by the ADA to fight discrimination. I understand that the subcommittee has much in the way of important matters before it this Congress, but I feel Mr. Graves' case deserves review. My office can provide volumes of documentation to support Mr. Graves' claims of discrimination and collusion against him. Should you have further questions regarding Mr. Graves or his case, please do not hesitate to contact Charles McCrudden of my Washington staff at 225-5261.

Respectfully,

James A. Traficant, Jr.
Member of Congress

JAT/cm

cc: The Honorable John Mica
Subcommittee on Civil Service

December 15, 1998

Stephanie Tubbs Jones
U.S. Congresswoman-elect
Justice Center
1200 Ontario
Cleveland, OH 44113

RE: AIDS BIOENGINEERING

Dear Congresswoman-elect Tubbs-Jones:

I recently contacted your office and learned you were away in Washington for your Congressional orientation. One of your fellow county prosecutors suggested I send you a letter with my concerns over the Pentagon's bioengineering of the AIDS virus. Your fellow prosecutor expressed the idea that the AIDS virus had not been man made but was accidentally placed in the hepatitis B vaccine trials sponsored by the Center for Disease Control and administered by the New York Blood Center in Manhattan.

Your fellow prosecutor believed that men recruited for the development of the vaccine may have had HIV antibodies in their blood, and this led to the subsequent development of a tainted vaccine. For the record, all of the more than 10,000 blood samples were subsequently tested for HIV. No incidence of HIV was found. Conversely, of the 1083 gay men who received the experimental vaccine in 1978, a large percentage of the stored blood samples subsequently tested positive for HIV antibodies. The HIV virus did not exist in the American population prior to 1979. I would be very happy to provide you and your colleague with the appropriate medical and scientific references.

In 1969, (July 1, 1969), the Pentagon asked Congress for $10,000,000 to create a 'synthetic biological agent'. For your convenience I am enclosing page 129 of U.S. House Bill 15020. I am further directing your attention to numbered paragraph 2. This evidence was submitted on September 28, 1998 in the federal court here in Cleveland. The case number is 98 CV 2209 (Graves v. Cohen). Judge Lesley Wells continues to sit on her hands, perhaps because of political persuasion or negative disposition toward people with HIV/AIDS.

I have contacted Representatives Clyburn and Green, with respect to being given an opportunity to present a full presentation of the information I have amassed to the Congressional Black Caucus.

I am seeking your support in this endeavor.

Sincerely,

Boyd E. Graves
1008 Elbon Road
Cleveland, OH 44121-1429
216.382.9252

enclosure

cc: Michael R. White, Mayor

February 15, 1999

TO: WILLIAM JEFFERSON CLINTON, PRESIDENT OF THE UNITED STATES OF AMERICA

SUBJECT: DEMAND FOR APOLOGY AND REPARATIONS FOR THE CREATION PRODUCTION AND PROLIFERATION OF THE HIV ENZYME

Dear Mr. President:

As a consequence of the abridgment of the national press and media, I write today, this 'Presidents' Day', to demand a resolute apology for the creation, production and proliferation of the human immunodeficiency virus (herein referred; the "SPECIAL VIRUS"). It is important to note that Dr. Gallo has facilitated the derivative of names, beginning first with the name: Human T-cell Leukemia/Lymphoma virus. Then Dr. Gallo changed the name to Human T-cell lymphotropic virus and subsequently to HIV. The experts believe the name change was a significant step in the process of disinformation to the world public. In actuality, the special virus is a contagious cancer that is the bi-product of our "United States Special Virus cancer Program" which began officially in 1962. It is worthy to note, our government, as well as others around the world, have persisted in venues of population control and eugenics, since the turn of the century.

The Congressional Black Caucus has committed to a de novo review of the supportive evidence, however, it is the full U. S. Congress that must prioritize this matter. The U. S. Congress with your fervent leadership must steadfastly pursue the re-education of the American people by publicly "exposing" and rapidly and deliberately "closing" this matter. The issue of the iatrogenic origin and development of the "special virus" is 'unparalleled' in recorded history, superseding the invention of the wheel and the discovery of fire. Our national security, increasingly intertwined with that of the world's, requires your immediate attention.

These demands are presented on behalf of Americans who have suffered and/or been killed as a direct result of our "special virus" federal program. Under separate cover, I will alert the Secretary General of the United Nations.

I write today in my capacity as lead plaintiff in the federal lawsuit for reparations and damages. As you are so intimately aware, we are a strong nation and a nation of laws. Accordingly, it is conceptually possible to genuinely forgive our scientists and the national press. Let the people of the United States, **together** with our scientists and press, consider our ever evolving science of human gene manipulation. If we allow this program to continue in secret, none of us will survive the inevitable darkness affiliated with the extinction of the human race. It is exceedingly clear our national press has been abridged and our Constitution infringed. We are appropriately enacting the available processes incumbent in this demand for an immediate Executive Order.

Our American history is replete with repayment and apology to its citizens for American violations of human and constitutional rights. Most recently, we have placed HIV Positive hemophiliacs in this class of persons. Previously, we have included the participants of the Tuskegee study. The ensuing synopsis of evidence presented herein conclusively proves a right to reparations for people with HIV/AIDS. We are a class of Americans who have experienced physical, psychological, economical and societal demise. But for the intentional, deliberate and intensive efforts of our government, we would not be so encumbered.

Fifteen years ago, the American Association for the Advancement of Science accurately concluded the government's "special virus" **(HIV) 'is a new pathogenic agent produced via human manipulation of genes'**. Following the Association's conference, foreign scientists returned home and informed their leaders of the "new" U.S. virus. One need look no further than a speech given by then South African President, P. W. Botha on August 18, 1985:

> **...Our Combat Unity is now training *special* White girls in the use of slow poisoning drugs. Ours is not a war we can use the atomic bomb to destroy the Blacks, so we must use our intelligence to effect this. The person-to-person encounter can be very effective. As the records show that the Black man is dying to go to bed with the White woman, here is our unique opportunity...We have received a new supply of prostitutes from Europe and America...**

Mr. President, these are the words of a **minority** White government. The tenor of his complete remarks and the "scientific secrets" claimed to be held with the United States demand a further expose of the true aims and objectives of the 'special virus" program. Further, the recently declassified National Security Defense Memorandum #314 (NSDM #314) leaves little doubt as to our country's secret and aggressive agenda of depopulation. Henry Kissinger Memorandum #45 is consistent with a **majority** White government depopulation scheme. Our aims, in support of our national security, require immediate concern and intervention. I respectfully request that my prior articulations to the Cleveland Plain Dealer and the New York Times newspapers, be incorporated into this request for immediate presidential intervention. Based on the following facts, it is eminently clear Americans, as well as others, have been killed, maimed and socially raped by an intentionally vicious depopulation and eugenics agenda of the United States of America:

INDISPUTABLE FACTS AS EVIDENCE OF DEPOPULATION:

1. The "special virus " program began officially in 1962, Litton Bionetics Inc., with eventual oversight by Dr. Robert Gallo. Infant primates were injected with an "immunosupressive" enzyme to study the leukemia process in humans. Each year, Litton Bionetics released surviving primates back into the wilds of Central Africa.

2. Dr. John Landon, Dr. David Valerio and Dr. Robert Ting were the project directors for the aforementioned.

3. Although the "special virus" program began in 1962 and annual reports were prepared by the National Institute of Health, none of the first seven progress reports exist in the public domain.

4. "Special virus" Progress Report #8, dated, August, 1971 (and subsequent ones) exists in 'very limited' fashion. According to Dr. Leonard Horowitz, the library of the National Institute of Health

(NIH) does not maintain the history of the progress of the program.

5. The "special virus" program is not catalogued in many of the libraries of agencies and universities that participated in the program.

6. Dr. Robert C. Gallo isolated the HIV enzyme in accordance with the "special virus" program, contract #: **NIH-71-2025.**

7. Dr. Robert C. Gallo, et. al., has systematically, intentionally and purposefully sought to thwart, disrupt, redirect and retard every meaningful effort leading to public discovery of his 1960's gene manipulation work.

8. The HIV enzyme was isolated and **mass produced** via illegal "roller bottle" techniques in association with "special virus" program, contract #: **NIH-NCI-E-70-2048.**

9. According to Dr. Horowitz, it is possible to grow 2.1 trillion virus particles per roller bottle batch!

10. The mass produced HIV enzyme was attached to the Hepatitis B experimental vaccine-Batch #751.

11. Hepatitis B experimental vaccine-Batch #751 was administered to 1243 men in Manhattan, New York in 1978 and 1979.

12. Part of the disinformation campaign of Dr. Gallo, et. al. provided that human subject experimentation was necessary to determine; a) symptomology, b) infectivity and c) lethality of the HIV enzyme.

13. The genome of the HIV enzyme is **identical** to the recombinant bovine visna virus, according to all scientific criteria for the identification of virus.

14. Since the late 1930's the U. S. National Academy of Sciences has served as the stealth backbone of the United States' depopulation program.

15. On June 9, 1969, the U.S. Pentagon informed the U.S. Congress of its offensive experiments which would lead to worldwide scourge and black death type plague.

16. On September 16, 1975, the Central Intelligence Agency informed the U.S. Congress of its 1967 creation of a 'poison dart gun' (microbioinoculator).

17. The CIA's poison dart gun can convey the "special virus" (or other biological agents) to a person or animal from a distance of 100 yards without observation by anyone including the intended victim(s).

18. An independent review of the more than 22 million blood samples maintained at the Army Navy Serum Repository (ANSR), the Walter Reed Army Institute of Research (WRAIR), the United States Army Medical Research Institute of Infectious Disease (USAMRIID) and the Henry M. Jackson Plasma/Cell Reagent Bank will definitively show the clandestine introduction of synthetic biological agents in the United States.

19. On July 1, 1969, the U.S. Pentagon informed the U.S. Congress of its plans for a 'synthetic biological agent'.

WHEREFORE; pursuant to the nonexclusive foregoing factual information I seek your individual support for an apology and an immediate independent review assisted by the United States Congress on behalf of all the good people of the United States. Contrary to the belief system of your close, personal friend and advisor, we the people are entitled to your "close" supervision in which to curtail this intentional genocide. THIS WORLD awaits your unadulterated leadership.

Respectfully submitted,

Boyd E. Graves,
lead plaintiff for AIDS reparations

cc:

On January 20, 2000 US attorney Lisa Hammonds Johnson notified lead plaintiff for global AIDS apology, Boyd E. Graves that the U.S Department of Justice had replaced the Department of Defense as "true defendant," naming the President of the United States. The court's action followed Dr. Graves' submission of the 1971 Special Virus Flow Chart as evidence to the Sixth Circuit in 1999. On April 30, 2000 President Clinton declared AIDS a threat to National Security and in an unprecedented move placed the National Security Council in charge of a Special Virus, AIDS. *Boyd E. Graves v The President of the United States* Case No 99-4476 would stagnate in the court over eight months before being silently dismissed under the cover of a presidential election November 7, 2000. Dr. Graves never received a written response to his 1999 President's Day letter.

January 5, 1999

The Honorable James E. Clyburn
United States Congress
319 Cannon HOB
Washington, D.C. 20515

Dear Representative Clyburn:

I would like to make a presentation to the Congressional Black Caucus on the information I have amassed which proves the Pentagon's involvement in the creation of the AIDS virus. Chief amongst the evidence is the enclosed sworn testimony of Dr. Donald MacArthur. Dr. MacArthur testified before the Subcommittee on Appropriations on July 1, 1969.

I have spoken with Yelberton Watkins of your staff, as well as the staff of Congresswoman Green.

I have been researching this matter for the last seven years. I am African American and I have AIDS. I also hold a law degree. I believe it is necessary to identify the true origins of the virus to hasten the process for a cure or vaccine.

I look forward to working with you and your staff and the Congress.

Sincerely,

Boyd E. Graves
1008 Elbon Road
Cleveland, OH 44121
216-382-9252

enclosures

cc: William Jefferson Clinton, President
c/o Susan J. Smith, Director, Agency Liaison Office
The White House

Donna Christian Green, Congresswoman

January 8, 1999

John Mangels
The Cleveland Plain Dealer
1801 Superior Avenue
Cleveland, OH 44114

Dear Mr. Mangels:

Pursuant to our conversation I am providing you the scientific and medical references in furtherance of the federal lawsuit (GRAVES V. COHEN 98CV2209). You genuinely seemed surprised that Judge Wells, on October 28, 1998, sought to dismiss the action against the Pentagon. Under section 1915, Judge Wells believed no set of facts exist which impart any degree of culpability to the Pentagon. On November 2, 1998, I filed a motion for reconsideration, essentially asking the judge to review U.S. House Bill 15090, page 129. I also provided her additional supportive documentation with respect to the factual and chronological timelines that have been developed. Since the motion for reconsideration did not require 'responsive pleadings', it is inexcusable and suspect as to why Judge Wells has not ruled. The rule of law is being violated right under the nose of one of America's top newspapers. The only issue pending before Judge Wells is whether or not this matter is frivolous. I am today, awaiting a conference call from the Honorable George White, Chief Judge.The American people, and particularly those impacted by the AIDS virus, are entitled to an expeditious disposition of this matter. The factual literature concludes the following: (1) Human AIDS virus not from monkeys & (2) The AIDS "epidemic" did not start in Africa.

In support of the foregoing I am suggesting you review a June 2, 1988 article from the **Los Angeles Times**, " RESEARCH REFUTES GREEN MONKEY THEORY". Please also see **Nature** 333: 396, 1988, Mulder, C. "HUMAN AIDS VIRUS NOT FROM MONKEYS". Please also see March, 1996 **Scientific American** ('No Start in Africa'), **Lancet** I: May 24, 1986, p.1217 Carswell, et al "HOW LONG HAS THE AIDS VIRUS BEEN IN UGANDA?" and **The London Times**, May 11, 1987, "SMALL-POX VACCINE TRIGGERED AIDS VIRUS".

Additionally, I have previously provided you with a **New York Times** article which questioned the existence of AIDS in the 1950's (April 4, 1995, "EARLIEST AIDS CASE CALLED INTO DOUBT").

The research reveals the 'appearance, structure and molecular weight of the virus indicate the AIDS virus is a genetic recombinant of two viruses spliced together'. See, inter alia, The Strecker Memorandum. For the record the two viruses that were spliced together are the bovine leukemia virus (cow) and the visna virus (sheep); ultimately forming the bovine visna virus (BVV). The BVV when "introduced" in humans, intelligently seeks out T-cells and destroys them, leaving the human body without an immune system. The BVV is a synthetic biological agent. This agent is consistent with the production timeline developed by the Pentagon as is represented in their July 1, 1969 testimony before the U.S House of Representatives, Subcommittee on Appropriations, U.S. House Bill 15090, page 129.

Sincerely,

Boyd E. Graves
Cleveland, OH 44121-1429

"HYPOCRISY, IGNORANCE AND SECRECY:
An Open Letter to the Cleveland Plain Dealer"

January 20, 1999

This letter is written in response to your cover story, **"Scientists try to counter bioweapons"**, Mangels, J., 11/23/98. The ensuing article cleverly danced around the most devastating biological agent known to mankind, the AIDS virus. The overwhelming factual evidence reveals an iatrogenic origin of the virus, firmly rooted within the military's offensive biological warfare program. Specifically, on July 1, 1969, the United States Pentagon sought appropriations from the United States Congress for the purposes of creating the virus. The AIDS virus is a 'chimera' consisting of two animal viruses splices together in a human culture. The AIDS virus is a product of recombinant (gene shuffling) DNA technology that flourished in the late '60's and is standard practice today in university and commercial laboratories around the world. The direct evidence of the military's involvement (in the splicing of the AIDS virus) lies in the sworn testimony given by the Pentagon on June 9, 1969 and July 1, 1969. The military's strategies of offensive and defensive biological warfare, have remained inextricably interwoven since the official inception of the government's biological warfare program in 1943. The ten year time frame, predicted by the Pentagon in 1969, for the onset of the AIDS epidemic, was realized with the first report of AIDS in America in 1979. That first reported case was a "cohort" (study participant) of the Hepatitis B experimental vaccine trials that began in Manhattan, NY in1978. The subsequent epidemiological profile of the early AIDS virus is an **identical match** to the cohorts of the Hepatitis B experimental vaccine trials. An intentionally contaminated vaccine trial propagated by the federal government. There is but one conclusion; a large number of persons were simultaneously "introduced" to the government's new infectious microorganism. In essence, a retrovirus bioweapon designed to debilitate and kill. This fact is further underscored when one considers the additional fact that the virus "introduced itself" to Africans and Americans at the same time. Large numbers of Africans were exposed via the small pox vaccines our government sent to Central Africa, and a large number of "underclass" Americans were exposed via the hepatitis B experimental vaccine.

Your article grossly and blatantly ignores the fact that our government has continued human subject experiments, in the name of military preparedness, for nearly seventy years. Many are aware of the syphilis experiments that began in earnest in 1932, and some are aware of radiation experiments that were conducted as a part of our atomic energy program. Few are aware of the LSD experiments, where victims received over seven million dollars, and even fewer are aware of the biological experiments aimed at African Americans in 1951. Offensive biological agents have been a mainstay of our military's preparation for more than a half century, and it is ludicrous to believe it does not continue. Relatedly, It seems suspiciously peculiar that only American troops suffer from Gulf War Syndrome. Your continued failure to espouse the factual history with respect to the AIDS bioweapon makes you susceptible to ridicule, and demeans the high standard of journalism your paper has maintained in the past. The obfuscation of your assistant editors will not silence the growing outcry for factual journalism, particularly when you possess credible and verifiable information, and conveniently fail to honor the true nature of your existence. Hypocrisy, ignorance and secrecy are truly our most feared biological weapons.

Sincerely,

Boyd E. Graves, J.D.
Director AIDS- Concerns Common Cause Medical Research Foundation

January 30, 1999

Phillip Bossey, Editor
The NEW YORK TIMES
229 West 43rd Street
New York, NY 10036-3959

Re: An Open Letter to the New York Times

"The AIDS Pandemic: Ancestral or Iatrogenic Origin?"

Dear Mr. Bossey:

I write in response to your article of December 27, 1998 entitled: **"AIDS in Africa: The Silent Stalker",** by Donald G. McNeil, Jr. I find it personally disheartening that you continue the gut wrenching practice of displaying photos of malnourished or debilitated Africans, to enhance the veracity or believability of your story. The photo purposefully preconditions your readers to a posture of sympathetic acceptance and elevates belief and conjecture to absolute fact. Even AIDS virus "discover" Robert C. Gallo defers to the newspaper for resolution of scientific uncertainty (the **New York Native** 9/9/84). My name is Boyd E. Graves. I am a graduate of Annapolis and law school. I am African American and homosexual. I received a diagnosis of AIDS on July 31, 1998. I am the lead plaintiff in the federal lawsuit for reparations for people with HIV/AIDS (GRAVES V. WILLIAM COHEN, 98 CV 2209, U. S. District Court, Cleveland, Ohio).

My initial review of the proffered information in Mr. McNeil's article, coupled with my current factual conclusions to the contrary, require that I present a rebuttal to your story. Initially, let me say that, I personally believe it to be important, and a high priority matter, that we unequivocally identify the origin of the AIDS epidemic. The 1976 African infection rate, as mentioned in your article, (5 people out of 454) does not support an 'ancestral' origin of the virus. This threshold point is further buffered by the Center for Disease Control's ("CDC's") 1986 African study, which revealed a similar, scant infection rate. There is a genuine dispute as to the true nature and origin of the AIDS microorganism. Identifying the true origin of the AIDS epidemic is of utmost concern, in that it might lead to greater therapeutic processes and/or quicker routes to a vaccine and/or cure. However, if that identification leads to our own government, and there is clear and convincing evidence of culpability; then, we the people, are entitled to a complete Congressional review and report. Congresswoman Donna Christian Christensen, Health Brain Trust for the Congressional Black Caucus ("CBC") has committed to the review process. According to the staff of CBC Chairman, Representative Clyburn, the CBC will adhere to Representative Christensen's recommendation. Our government does have a well established litany of culpability of numerous instances involving individual/human rights violations. We also have established procedures to compensate non-consentual victims. The HIV/AIDS populace are unjustly afflicted and as such are entitled to economic reparation, at whatever the cost. This letter is written, with respect and concern, to awake and provoke the sleeping tsunami of public outrage. This letter is written as a direct challenge to the intentionally misleading and bogus information (standard government AIDS propaganda) contained in your article. If our top scientists (like Dr. Gallo) rely on the print media for answers, direction and fact, then you have significant collateral responsibilities to the American people and to the people of the world. Your newspaper must ultimately determine, and by your integrity, prove, that you are not a vehicle for dogmatic, hypothetical prognostication. Your newspaper must stand and declare that you are truly the people's foreman. I am also hopeful, that your newspaper will indulge me, and provide a subsequent opportunity to make an unfettered presentation. I am exceedingly grateful to a number of "man-made AIDS" pronouncers who have preceded me; particularly, (the late) Theodore A. Strecker, M.D., (the late) Zears L. Miles, Jr., (the late) Douglas Huff, Eric Taylor, Alan R. Cantwell, M.D., Leonard G. Horowitz, D.M.D., M.A., M.P.H. and Robert R. Strecker, M.D., Ph.D. I am truly honored to walk in the shadows of their footsteps and I will exalt to keep the message sharply focused. On the strength of my individual resolve, (and in light of our government's checkered past), I offer that it is neither overly suspicious or delusional to be certain of the iatrogenic (man-made) origin of the AIDS epidemic.

Historical Perspective.

The United States' retrovirology (AIDS-like viruses) program began "officially" in 1943. However in 1941, the National Academy of Sciences appointed a special committee, who subsequently recommended research on the offensive potential of bacterial weapons. The U. S. retrovirology program selected George W. Merck, by default, to head the federal program, deceptively disguised as the War Research Service. In 1943, President Roosevelt went public and announced the creation of our biowarfare program as "The Birth of Science", and appropriated funding for the creation of a research facility, Camp (later Fort) Detrick in Fredrick, Maryland. According to President Roosevelt, **Ft. Detrick was established for two purposes; a) develop defensive mechanisms against biological attack, and b) develop offensive biological agents to respond "in kind", if biologically attacked.** In a self-serving statement in 1947 (**"Biowarfare Investigations of 12 December 1947"**), our government justifies the necessity to conduct human subject biowarfare experiments. Our government defended the amnesty deal made with captured Japanese biowarfare scientists. We agreed to 'not turn them over to the Soviets, so long as the United States could continue the Japanese human subjects biowarfare research'. The Japanese had begun their human subject retrovirology experiments during the early 1930's, while, at the same time, the United States was just beginning its secret virological experiments (syphilis). The United States secretly incorporated and adapted the Japanese biowarfare technology into its growing biological research on recombinant retrovirology. We also maintained an "active" program after the war. Apparently, so did the Soviets. A review of a boastful 1960 speech by Soviet Prime Minister Nikita Krushchev affirms the Soviets had an equally advancing retrovirology program:

> **We have a *new* weapon, just within the portfolio of our scientists, so to speak— so powerful that, if unrestrainedly used it could wipe out all life on earth.**

Minister Krushchev's comment is hauntingly surreal and our government had legitimate national security reasons (in the 50's) to continue to explore and assimilate theaters of biological war. Accordingly, our earliest preparations reveal a first biological warfare drill in 1951. There is sufficient proof we conducted a "placebo" biowarfare virus attack aimed at African Americans. Naval supply depots were selected, 'solely because a large number of crate handlers were Negro'. The 'placebo' biowarfare agent was a derivative of the "Valley Fever" virus. Valley Fever has a clear, identifiable epidemiological fingerprint of an affinity toward African Americans. It is indisputable that African Americans are ten times more likely to contract Valley Fever than are Caucasian Americans. Thus, any experiment to test a biowarfare agent containing antigens of Valley Fever, is an experiment targeting African Americans. The specific records of the 1951 biowarfare attack were declassified in 1981. It is currently unresolved as to there being **any** biological similarities between African American and Russians. It is indisputable our government has conducted virological (share croppers) and retrovirological (Philadelphia) attacks aimed at African American citizens. The AIDS virus is a racially motivated agent of depopulation in that it preys upon the "OKT4 epitope deficiency" indigenous to the Black population. The American government has a pattern and practice of directing offensive biological research at particular racial or ethnic groups of its own citizenry. This practice of 'singling out citizens', in this case, African Americans, is inconsistent with the biowarfare purposes defined by President Roosevelt in 1943. **It is simply inconceivable that the United States would face a biological attack from African Americans**, and be forced to respond "in kind". Therefore, one can only conclude, our government continues to focus and pursue a racially biased biowarfare research agenda, directed at its African American citizenry and the Black populations of the world. In a "worst case" AIDS scenario, the virus will kill all Blacks, but only 95% of all whites. It is conceivable to believe and worthy of public attention, that our government has conducted biowarfare on other African populations. It is more than hypothetical hyperbole to conclude that our government has conducted biowarfare on Black Africa. It is fact.

Iatrogenic origin.

History will inevitably show the AIDS pandemic to be the single most devastating event in the history of mankind. Oddly enough, a biohazard epidemic (worldwide scourge) was of such concern in the late 1940's, that a secret biowarfare conference was held in Ottawa, Canada in 1952. The United States cannot deny its participation in this early, human depopulation conference. The United States participated in a similar conference in Lyons, France in 1964. The participation in these conferences, as well as others, is further credible evidence of an "active" population control agenda, which ultimately appears to have "Black and homosexual cross hairs"! Thus, in somewhat elementary deference to your newspaper, it is most likely there are stored blood samples, from the 1950's, that evidence our government's early work in recombinant retrovirology or that of another government's. Clearly our history of offensive biological warfare in the Korean War is a matter of documented record. The new facility built at Ft. Detrick in 1977, as mentioned in your article, was not for the purposes of "creating" a new infectious microorganism. The new facility was more for the doomed research that would attempt to re-bottle a genie that had been intentionally constructed and set free. Your hollow and deflective assault on errant Soviet propaganda is severely undercut by the outcries of the 'total' international scientific community. They are not all simply communists, lest we all be ourselves, and I am not (a communist)! In a 1974 desperate attempt (**Science**, 7/26/74) to salvage the government's illusion of a non-existent theater of population thinning, the National Academy of Sciences-National Research Council ("NAS-NRC") issued a public declaration calling for a moratorium on recombinant retrovirology research and experiments:

> **Scientists throughout the world (should) join with the members of this committee in voluntarily deferring experiments (linking) segments of the DNA from (cancer-causing) animal viruses to... possibly increase the incidence of cancer or other diseases.**

This hapless pronouncement attempted to "freeze" national and international research concerning recombinant retrovirology. In other words, our government knew it could not get the genie back into the bottle. They thought 'freezing it' might be an effective means of controlling the microorganism and extending the delay of the unavoidable "returning boomerang". The government's pronouncement was a flag of surrender to the superior adaptability and fighting skills of a "new" microorganism it created. According to the **Cleveland Plain Dealer**, Mangels, J., **"Scientists try to Counter Bioweapons"**, 11/23/98, 'the Pentagon is supporting research on biological warfare in scores of federal, private and collegiate laboratories around the world'. However, many top government and military retrovirologists believe there is no adequate biological defense against the "synthetic" microorganism.

Scientific Probability.

Mr. McNeil's article appears to intentionally and purposefully ignore, expressed mathematical epidemiological improbabilities of any association of the AIDS pandemic and 'urban migration patterns' in Central Africa. Mr. McNeil failed to inform your readers that the CDC in 1986, (Drs. McCormick and Fisher-Hoch), found a similar rate of infection in Africa to that which existed in 1976 (5 Africans out of 454). Equally, there is no epidemiological fingerprint of an evolutionary derivative of the pandemic associated with the African continent. In 1986, **The Lancet** published the results of a study involving elderly Ugandans. None tested positive for HIV. **The mathematical epidemiological profile of the pandemic does confirm however, that large numbers of Black (not White) Africans and homosexual Americans were (in the 70's) simultaneously introduced to a "new" infectious microorganism.** I want to reassert at this point, the high probability and likelihood, that early blood samples, when retested, **should** carry the potential to be positive for HIV antibodies. In light of the world's predilection for retrovirology, it was only a matter of time before there would be definitive proof of population control and exacting human subject experimentations. This factual discovery is not inconsistent with, perhaps surprisingly to Mr. McNeil, the notion the AIDS pandemic has an iatrogenic (man-made) origin. It is important to recall that we have had an active retrovirology program since 1943. In essence, the realization of this dark, factual revelation is an intellectual leap of a caliber equivalent to "grade school hopscotch". It is rational to conclude our government's population control agenda includes secret, human subject inoculations. Today, many nations and governments are well-versed in the processes of microbiology, gene splicing and recombinant retrovirology technology. I call upon your newspaper to seek to make public the agenda, papers and records of the 1952 secret retrovirology conference in Ottawa, and the 1964 conference in Lyons, France. Additionally, and for the record, your newspaper has an obligation to seek to make public the agenda, papers and records of our own government's controversial retrovirology conference; "Entry and Control of Foreign Nucleic Acid", April 4 -5, 1969. This conference was hosted by the American Institute of Biological Sciences at Ft. Detrick. This conference was met with international protest and scientific boycott. Why? Metaphorically speaking; we were advertising that 'something interesting has come out of our bottle, let us see what it will do'. Many "moral" scientists wanted nothing to do with our government's genocidal genie bottle. The genie turned out to be a contagious "new" microorganism capable of inducing immunodepression by affecting T cell function. Our government had 'Frankensteined' a contagious cancer. Several months later, the Pentagon was ready to mass produce and mass microbioinoculate.

Military, CIA and WHO involvement.

Accordingly, it is worthy of further interest to note, that George W. Merck, our country's top retrovirologist, was also a collaborator with the Central Intelligence Agency ("CIA"). The CIA's Projects "MKNAOMI" and "MKULTRA"(beginning in the 1950's), were the purveyors of clandestine and imperceptible inoculations in the United States, Central Africa and other countries. They exist for the purposes of population, mind control and region destabilization. This incriminating fact is further underscored when one considers that it is statistically impossible for a remote area of Africa to yield two genetically different "new" viruses (AIDS and Ebola) at the same time. Relatedly, it is also statistically impossible, for the AIDS epidemic to originate, at the same time, on two non-contiguous continents; and **needlesstosay**, two diametrically opposite populations! These are the indisputable pieces of evidence which point very strongly to an iatrogenic origin (CIA INJECTION) of both viruses. I call upon your newspaper to seek to make public, all CIA activities connected with the two CIA projects and the CIA's microbioinoculator (BIOLOGICAL DART GUN). **Former CIA Director, William S. Colby testified to the existence of the CIA's microbioinoculator in 1975, before a special Senate Select Committee.** In short, our government, (Ft. Detrick, NCI, NIH, CDC, NAS-NRC and the CIA), in conjunction with the World Health Organization ("WHO") were in total conspiratorial agreement with their seething desire to realize "population thinning". <u>This</u> is the reason our April 4-5, 1969 conference was met with repugnant indignation from truly moral scientists from here and around the world. However, this indomitable force, our government and the WHO, with the blessings of the Council on Foreign Relations ("CFR"), easily recruited and infiltrated pharmaceutical companies, and began 'thinning the population' in earnest, in 1967, starting with Central Africa. It is peculiar to note, that one of the stated purposes of the WHO is the development of prophylaxis measures and mass immunizations. How does your newspaper explain away the WHO's 1950's -60's research centered on "slow-acting" viruses? These slow-acting viruses are in a group of viruses called "lentiviruses". As a supportive point of information, AIDS is classified as a member of the lentivirus group in that, there is a somewhat long time frame between infection and noticeable illness. This is the characterization of AIDS and AIDS-like viruses that make them ineffective as military weapons. In other words, it would not make sense, 'in the heat of battle', to deploy a recombinant retrovirus, 'as a military weapon'. The effect would have no immediate impact on the battle at hand, and as such, would not provide any advantages to a nation at war. Our government expressed this exact point on June 9, 1969 before the Subcommittee on Appropriations. **The government's testimony of June 9, and July 1, 1969 leaves no doubt of the government's involvement in the creation, production and proliferation of an AIDS-like virus. <u>The government's testimony confirms they were working on the creation of a specific "new" infectious microorganism, 'one that does not naturally exist and one for which no natural immunity could have been acquired'</u>.** Twelve years later, the word AIDS would be created to give a name to this "new" microorganism. A further review of the June 9, 1969 testimony reveals <u>conclusive</u> <u>proof</u> of our government's deployment of offensive biological agents capable of achieving population thinning by inducing a worldwide scourge and black death-type plague. The release of the government's **"synthetic"** biological agent coincides with the epidemiological profiles of the small pox vaccine in Africa and the hepatitis B experimental vaccine in the United States. Every legitimate scientist in the world recognizes and concedes that some sort of <u>mass</u> infection (microbioinoculation) occurred in the 1970's. I find it notably peculiar that your article makes no reference whatsoever to the congressional testimony or the "strikingly similar" epidemiological profiles of the vaccines and the virus. However, in fairness, your AIDS article of March 3, 1987 does unwittingly confirm a higher HIV rate of infection in the United States than in Africa.

Conspiratorial journalism.

Unfortunately for the people, only two months later, you balked when given a chance to make a more definitive presentation. Specifically, your newspaper was given an opportunity in May, 1987 by the **<u>London Times</u>**, to investigate further a connection between the African AIDS epidemic and smallpox vaccine. For some unknown reason, your newspaper chose journalistic silence. Your newspaper chose to say nothing and, in so doing, provided further cover for the CIA's biological dart gun and their population thinning agenda. Additionally, your article of June 30, 1987 is more of a 'mystery' than the article itself. You gloss over your March '87 article and ignore the fact that, **<u>retested American blood samples had a higher infection rate than African samples!</u>** Yet, you continue to pontificate of an African ancestral origin of the epidemic. Your collaborative silence and seemingly intentional misdirections, impugn your professed standards of excellence and integrity. Your impugnation extends almost to a point where you have imploded under the weight of conventional belief and disinformation. Your April 8, 1997 article about Ft. Detrick led the world to believe that the 'fort' had stopped biological warfare development in 1969. Is there one amongst us who believes you? Your newspaper has a discernible, unscrupulous history of printing 'cover stories' for the CIA and military to divert the minds of the people. The fiasco with C.L. Sulzberger is a glaring example of your propensity to 'double-dip', to the continuing detriment of the American people. In light of the overwhelming factual information available to the contrary, your article of December 27th appears to be more of the same, as well as your article of Nov. 28, 1991. However, it is not only your purposeful act to ignore substantive evidence, of which you should be faulted, it is also your willingness to serve as the state's agent to alter the mind of the people.

Simply put, you are perhaps, unknowingly, fulfilling the objectives of population thinning by soothing the mind of the people. If you continue, your contributory atrocities will be unmatched ***in this century.***

The dangers of population thinning.

Your belief that the AIDS epidemic started in rural Africa is consistent only with our government's secret project, MKNAOMI, to place it there. The AIDS epidemic is not one rooted in ancestral ilk of famine or disease. Of course, the photo associated with your December 27th article does reinforce your propaganda of a biological cesspool, even though it is we who are filling the pool! The western world has a long and tattered history of unsavory biological (and chemical) experiments on poor countries and their people, **especially** mineral-rich Africa. It should not be surprising that our government has participated in research of offensive biological agents that are man-made and uncontrollable. An independent, third party review of the stored blood at the Army-Navy Serum Repository ("ANSR") in Rockville, MD would easily identify which agents are natural or synthetic. Synthetic biological agents, like AIDS, are officially classified as uncontrollable, because once they are "introduced" into human populations, even into remote populations, there is no adequate way to safeguard the initiator's own population. The "boomerang" effect of microbiology capitalizes upon the microorganism's 'failure' to localize geographically or ethnically. In spite of extensive knowledge and acute awareness of the potential catastrophic danger, our government made clear, conscious decisions to participate in population thinning. Which raises an interesting tangential point of inclusive discussion. Is there a juxtaposition between the agendas of the "Pro-Life" movement and Population Thinning? In other words, is the Pro-Life movement merely a "smoke screen of ideology" to shroud and conceal, the agendas of population thinning and eugenics? Speaking abstractly, this translates to: "Every baby must be brought to term, it is we (w.h.o.) kill. Go forth and multiply, it is we (w.h.o.) divide." Where is the outrage of the Pro-Life movement?

With the collaborative support of the WHO and CFR, it is clear we have decided to continue to proceed with the process of population thinning. The population thinning agenda calls for our country to have an 'ideal' population of 100 million. They would then be encourages to procreate unabashedly. Black Africans and homosexuals were initially selected to cloak the population thinning scheme and "to avoid serious legal and logistical problems". Your newspaper has continually and conveniently failed to print the truth. It is painfully true our government has previously chosen African American sharecroppers and crate handlers for biowarfare, but the choice of homosexuals takes on an added culpable dimension. Not only were homosexuals chosen because of their low social pecking order, they were also chosen because they appeared to represent the best group, that could keep the microorganism "maximally localized", in their own community. Even as late as 1993, the NAS-NRC released a statement to the **Wall Street Journal** (3/17/93) to stem public hysteria and mask the impending 'boomerang' effect of the synthetic biological agent:

> **HIV infection and AIDS will remain limited to specific geographical areas and risk groups identified at the beginning of the epidemic: gay men and more particularly an ever-growing population of urban, drug-addicted, poverty-ridden, malnourished, hopeless and medically deprived people.**

By introducing the microorganism to the homosexual community, our government simply bought additional time in which to study the effects of its new "synthetic" recombinant retrovirus. By contaminating cohorts (promiscuous homosexuals) of the hepatitis B experimental vaccine, our government has willingly demonstrated, **again**, its overt intention and desire to inflict injury, harm, oppression and death on a segment of its own citizenry. Our government has always been willing to risk a few, to give a noteworthy appearance of doing something for the many. However, I believe your newspaper has an uncontroverted obligation, to unmask the diabolical nature of our retrovirology programs, and to make public the true nature of the AIDS epidemic. Recombinant retrovirology is no longer a secret anywhere in the world. The American people and indeed, the people of the world, are entitled to the knowledge of the specific good and bad acts of our government. These (bad) acts stand as the "self-inflicting" and "self-incriminating" catalytic building blocks, of the onset of the AIDS pandemic. The existing factual evidence stands in direct contradiction with your newspaper. Quite frankly, your newspaper does appear to be purposefully standing in the way, and purposefully clouding the minds of the American people.. The evidence does not support your suggestion of an epidemiological profile befitting African "urban migration" or "ancestral heredity". Your suggestions do not fit any mathematical model for the origin of the virus or the "explosive" transmission rate of the late 70's and 80's. Specifically, if a stored blood sample revealed the presence of AIDS in 1959, there is no 'doubling factor' mathematical scenario, that yields an infectious kill ratio consistent with the large numbers of infection worldwide. **There is but one conclusion, a large number of persons were simultaneously infected ("introduced")**. Even both "so-called discoverers" of the AIDS virus, Drs. Montagnier and Gallo, refuse to rule out an intentional "introduction" of the AIDS virus. How can your newspaper? This premise of an intentional introduction is further corroborated by June Goodfield (author of **"Quest for the Killers"**), who documented the 'genuine fears' of Dr. Wolf Szmuness. Dr. Szmuness conducted the hepatitis B experimental vaccine trials and was absolutely convinced of an intentional contamination (CIA microbioinoculation)of the hepatitis B experimental vaccine. Dr. Szmuness was so sure of the CIA's contamination that he stored all 13,000 blood samples relative to the experimental vaccine! Those American blood samples were ultimately tested and the results proved earth-shattering. AIDS did not exist in our country prior to 1978. Additionally, and for the record, the virus's modern day profile (African Americans) proves the virus has no physiological predisposition or preference for White homosexuals. To conclude otherwise, requires additional extensive psychosomatic submission (brain-washing).

A compelling case of an iatrogenic origin.

The overall impression of your newspaper and our government is more graphically illustrated by the classic movie, "The Wizard of Oz". If we 'pay no attention to the man behind the curtain', we will believe. However, just as it is in the movie, the current has become agape. It is absolutely false that Dr. Gallo discovered the AIDS virus in 1984. The medical and scientific literature reveals Dr. Gallo was already investigating "very similar" issues as early as 1971. The world is entitled to a cross examination of Dr. Gallo and Dr. Carlton Gadjusek, inter alia. A review of our government's congressional testimony of June and July, 1969 reveals that research on the "new" 'synthetic biological agent' had "officially" begun in 1967. The people are entitled to the facts concerning the true agenda of the NAS-NRC, going back to the 1930's. The government's testimony of June and July, 1969 further reveals that a group of experts had already considered the creation of a "new" synthetic (man-made) immunosupressive microorganism. They had considered a "new" microorganism capable of a chronic and specific decimation of the human immune system. The government's testimony is direct evidence of a man-made origin of the virus. A virus that was intentionally seeded for the purposes of achieving a genocidal holocaust of the disenfranchised.

Dr. Piot's credibility.

Even with the foregoing, it is still, yet for another reason that your December 27, 1998 article raises a deep and unsettling controversy deserving of additional factual investigation. Your article makes reference to Dr. Peter Piot, Executive Director of UNAIDS in Geneva, Switzerland. Dr. Piot's conclusions of an evolutionary AIDS virus are contradicted by his own prior statements. Dr. Piot has previously concluded the AIDS virus is man-made. Specifically, on April 8, 1998, and only days after he learned of the 1959 blood sample, Dr. Piot informed the **New York Amsterdam News** that 'the AIDS virus has a laboratory origin'. Dr. Piot commented:

> **The biological gap between the monkey virus and human host is too wide to be bridged in a single step; *even with direct injection of (monkey) blood (in humans)....***
>
> **I emphasized in a single step in the quotation above, because it takes many steps in a laboratory to engineer an animal virus to infect humans. AIDS is an animal virus, probably bovine visna virus, adapted for man. It was no accident.**

There is significant disparity in Dr. Piot's published comments! Consequently, there exists a meritorious scenario deserving of a greater degree of journalistic intrigue and responsibility. A sufficient challenge has been presented to the accuracy of a published article in your newspaper. Is the American public entitled to additional scrutiny by your newspaper? I believe the answer is yes. I also believe that, just as other human subject experiment victims have been "economically repaired" and compensated, so, too, should the American people, and quite frankly, all the people of the world, who have been injured or killed. Our government allowed our search for a cure for cancer to careen uncontrollably. We are to blame for the development and deployment of the "new" biological offensive agent which has created worldwide scourge and black death-type plague. The current federal lawsuit for reparations for people affected by HIV and AIDS is meritorious and factually compelling. However, because of prevailing conventions of thought, the federal judge has initially determined this to be a frivolous matter. I sincerely hope this letter serves to further alert the judge and the world, there exists a genuine factual dispute for which a jury of the people is required. The compilation of the direct evidence (the Congressional testimony) and the substantial circumstantial evidence(including the expert epidemiological profiles) will meet and exceed the highest legal standard for culpability. Each victim of the AIDS pandemic is innocent., whilst the opposite is true for our government. Just as the Congress has found American hemophiliacs to be deserving of reparations, so too are, in essence, **all other persons who were also intentionally infected.** Accordingly, you must seek to ascertain the true nature of 'Gulf War Syndrome'. Why do only American veterans suffer the affliction? Your newspaper has an obligation to draw the informed public's resolve to the assistance of a "listing social ship". It is demonstrably clear, we are in the midst of another human holocaust of astronomical proportion. We need your leadership. We cannot delay a single day more. It is now that we are called to speak. Our silent indifference is now abandoned. We will extinguish your cozy intransigence to correct disinformation. Your newspaper, has an opportunity, and an obligation to integrity, to right a wrongful course of humankind. Your newspaper can exorcise your dubious past, and place itself at the forefront of the mantel of truth, justice and credibility into the 21st century. This world awaits your response.

Sincerely,

Boyd E. Graves
lead plaintiff for AIDS reparations

DEPARTMENT OF DEFENSE

DIRECTORATE FOR FREEDOM OF INFORMATION AND SECURITY REVIEW
1155 DEFENSE PENTAGON
WASHINGTON, DC 20301-1155

09 FEB 1999

Ref: 99-F-0728

Mr. Boyd E. Graves
1008 Elbon Road
Cleveland, OH 44121-1429

Dear Mr. Graves:

This responds to your January 29, 1999, Freedom of Information Act (FOIA) request to the Secretary of Defense. The Office of the Secretary of Defense referred your request to this Directorate on February 3, 1999. The telephone conversation with Commander Voorhies of this Directorate on February 5, 1999, refers.

Please understand that this Directorate's holdings for any testimony before Congress does not go back to the 1969 timeframe. Records of that time frame have been destroyed in accordance with current records disposition schedules from the National Archives and Records Administration. However, we did locate the enclosed documents which are responsive to that portion of your request for Dr. MacArthur's July 1, 1969, testimony. We were unable to locate anything for the June 9, 1969, testimony. These documents were sent to this Directorate as enclosures to FOIA requests by other members of the public, which is why we were able to locate them. If full testimony of Dr. MacArthur exists, it would be in the Library of Congress' archives. You may call the Library of Congress Reference Referral Service at (202) 707-5522 for additional information.

Should you deem the no record portion of this response to be an adverse determination of your request, you may appeal this finding by offering justification to support an additional search effort. You should be aware that should the second search still not produce any records responsive to your request, you may be assessed fees pursuant to DoD Regulation 5400.7-R. Any such appeal should be forwarded within 60 calendar days of the date above to this office.

There is no charge for processing your FOIA request in this instance.

Sincerely,

A. H. Passarella
Director

March 13, 1999

Robert C. Gallo
Institute of Human Virology
725 W. Lombard Street, Suite S-307
Baltimore, MD 21201

Dear Dr. Gallo:

Please contact me at your convenience. I am interested in discussing your work with respect to the "Special Virus" program. We have filed a federal law suit for AIDS reparations and believe your expert testimony will be essential to the completion of this matter.

You did not mention in your book your work as a "Project Officer" for the "Special Virus" program.

Thank you for your commitment to the American people.

Sincerely,

Boyd E. Graves,
lead plaintiff for AIDS reparations

cc: Sue Smith, Director
The White House
The Honorable Lesley Brooks Wells

CASE WESTERN RESERVE UNIVERSITY

UniversityHospitals
ofCleveland

DS CLINICAL TRIALS UNIT
vision of Infectious Diseases
partment of Medicine

May 25, 1999

Mr. Boyd Graves
1008 Elbon Road
Cleveland Heights, OH 44121-1429

Dear Mr. Boyd:

Thank you for sending me the brief report on the early Gallo studies. I am aware of this work but cannot perceive the relevance of this work to the concerns you expressed to me.

Best wishes,

Michael M. Lederman, M.D.
Professor of Medicine

ling address:
niversity Hospitals of Cleveland
ivision of Infectious Diseases/ACTU
)61 Cornell Road
leveland, Ohio 44106-5000
16) 844-8175
ax (216) 844-5523

MIKE DeWINE
OHIO
140 RUSSELL SENATE OFFICE BUILDING
(202) 224–2315
TDD: (202) 224–9921
senator_dewine@dewine.senate.gov
www.senate.gov/~dewine

United States Senate
WASHINGTON, DC 20510–3503

COMMITTEES:
JUDICIARY
CHAIRMAN, SUBCOMMITTEE ON ANTITRUST
HEALTH, EDUCATION, LABOR, AND PENSIONS
CHAIRMAN, SUBCOMMITTEE ON AGING
INTELLIGENCE

May 25, 1999

Boyd Graves
1008 Elbon Road
Cleveland, Ohio 44121-1429

Dear Boyd:

Thank you for contacting me regarding the Special Virus program. I appreciate your taking the time to inform me of your views.

It is important to understand how you and all Ohioans feel on matters of concern and how these issues impact your daily lives. Should any federal agency propose to take any action or should legislation come before the full Senate on this issue, I will be sure to keep your views in mind.

In the future, I would be glad to receive other ideas or concerns you may have. Should you wish to contact me again, please feel free to call or write any of the offices listed on this page. You may also reach me via e-mail (senator_dewine@dewine.senate.gov), or visit my home page on the Internet at: *http://www.senate.gov/~dewine.*

Again, thank you for bringing your concerns to my attention.

Very respectfully yours,

Mike

MIKE DeWINE
United States Senator

RMD/bjn

STATE OFFICES:

105 EAST FOURTH STREET ROOM 1515 CINCINNATI, OH 45202 (513) 763-8260	600 SUPERIOR AVENUE EAST ROOM 2450 CLEVELAND, OH 44114 (216) 522-7272	37 WEST BROAD STREET ROOM 970 (CASEWORK) COLUMBUS, OH 43215 (614) 469-6774	37 WEST BROAD STREET ROOM 960 COLUMBUS, OH 43215 (614) 469-6774	200 PUTNAM STREET ROOM 522 MARIETTA, OH 45750 (740) 373-2317	420 MADISON AVENUE ROOM 1101 TOLEDO, OH 43604 (419) 259-7536	265 SOUTH ALLISON AVENUE ROOM 105 XENIA, OH 45385 (937) 376-3080

PRINTED ON RECYCLED PAPER

GEORGE V. VOINOVICH
OHIO

317 HART SENATE OFFICE BUILDING
(202) 224-3353
TDD: (202) 224-6997
senator__voinovich@voinovich.senate.gov
www.senate.gov/~voinovich

United States Senate

WASHINGTON, DC 20510-3504

GOVERNMENTAL AFFAIRS
CHAIRMAN, SUBCOMMITTEE ON
OVERSIGHT OF GOVERNMENT MANAGEMENT
RESTRUCTURING AND THE
DISTRICT OF COLUMBIA

ENVIRONMENT AND PUBLIC WORKS
CHAIRMAN, SUBCOMMITTEE ON
TRANSPORTATION AND INFRASTRUCTURE

SMALL BUSINESS

ETHICS

August 31, 1999

Boyd Graves
1008 Elbon Rd.
Cleveland, Ohio 44121

Dear Boyd:

Thank you for contacting me regarding your research on AIDS. I appreciate hearing your views on this important matter.

In your email you mentioned evidence that you uncovered, which points to a "hidden federal program" entitled "Special Virus." One of the most important aspects of my job is keeping informed about the views of my constituents. It is evident that you have thought a great deal about this topic, and I appreciate you taking the time to share your findings with me.

Please feel free to keep me abreast concerning further developments in your research.

Sincerely,

George V. Voinovich
United States Senator

GVV/PP

STATE OFFICES:
36 EAST 7TH STREET
ROOM 2615
CINCINNATI, OHIO 45202
(513) 684-3265

1240 EAST NINTH STREET
ROOM 2955
CLEVELAND, OHIO 44114
(216) 522-7095

37 WEST BROAD STREET
ROOM 970 (CASEWORK)
COLUMBUS, OHIO 43215
(614) 469-6774

37 WEST BROAD STREET
ROOM 960
COLUMBUS, OHIO 43215
(614) 469-6697

420 MADISON AVENUE
ROOM 1210
TOLEDO, OHIO 43604
(419) 259-3895

VIA EMAIL

September 8, 1999

Dear Senator Kennedy (via Scott Boule):

On August 21, 1999, in Canada, world scientists saw for the first time a 1971 U.S. flowchart that appears to spawn the AIDS virus. (Publisher's note: See - PHASE V "AIDS: MIRCLE IN CANADA" VHS 1999)

Dr. Garth Nicholson, CEO of the Institute for Molecular Medicine in Huntington Beach, CA could not contain himself and interrupted my presentation to "personally review" the flowchart. Dr. Nicholson's number is: 714-903-2900. Dr. Thomas Dorman has a copy of the flowchart. Dr. Dorman's number is: 253-854-4900. Chemist Vincent Gammill has a copy of the flowchart. Mr. Gammill's number is: 619-793-0293.

This matter should not continue one day further without you hearing from Dr. Len Horowitz: 888-508-4787 and Dr. Alan Cantwell: 323-462-0197. Have you turned a deaf ear to Professor Francis Boyle (217-333-7954), author of the U.S. Anti-Biological Terrioist Attack Act of 1989?

The 1971 AIDS Flowchart and the fifteen ensuing progress reports of the "Special Virus" program of the United States of America deserve immediate Congressional attention.

How long will it be before the American people learn what was reported in the 1984 Yearbook of the Stockholm International Peace Research Institute (SIPRI)?

How long will it be before the American people learn that AIDS is genetically linked to the German-made Visna virus?

How long will it be before the American people learn the only goal of the "Committee of 300" is depopulation?

In light of the flowchart and the progress reports, some of us are simply tired of the propaganda. The United States must openly admit its role in the culling of the human population. It must now be presupposed that the June 9, 1969 testimony of the Pentagon's Dr. MacArthur is accurate as to his truthfulness regarding the development of an offensive biological agent that would lead to world-wide scourge and black death type plague. In 1977, the "Special Virus" program produced 60,000 liters of a "new" immunesuppressing virus. Shortly thereafter we had noncontiguous epidemics.

The progress reports do reveal the secret as to why AIDS appears to have an affinity toward people of color.

We seek our right to question government, and we seek openess, in which to subsequently position ourselves to weigh forgiveness. It is time for the American people, and the people of the world to view and learn of this flowchart and its' accompanying progress reports.

I write today on behalf of the Constitution of the United States of America.

Freedom's slave,

Boyd E. Graves, JD
Director-AIDS CONCERNS
the Common Cause Foundation

September 14, 1999

TO: ROBERT G. ANDERSON
"60 MINUTES" —CBS TELEVISION

FM: BOYD E. GRAVES, JD
DIRECTOR-AIDS CONCERNS
THE COMMON CAUSE MEDICAL RESEARCH FOUNDATION

SUBJ: FACTUAL ERROR; YOUR 9/12/99 BROADCAST

RE: THE CONCLUSIVE EVIDENCE OF THE LABORATORY ORIGIN OF AIDS

Dear Mr. Anderson:

On Sunday, September 12, 1999, you were listed as the producer who concluded the lab origin of AIDS 'story' is without merit. We ask that your show do a retraction of the scoffing remarks you made about the laboratory origin of AIDS. A 1971 flowchart was recently presented in Canada and appears to represent the research logic of a "hidden" federal program that would create a prototype AIDS-like virus. Dr. Garth Nicolson was in Canada and has reviewed the flowchart.

However it is not the flowchart alone which definitively proves the origin of AIDS. The program was called "Special Virus" and it began in 1962. Apparently the collaborators met every year at Penn State as guests of Professor Fred Rapp. The program porduced fifteen (15) progress reports which correlate the experiments, contracts and contractors to the flowchart.

It appears the U.S. has destroyed the first seven reports. We have some of the remaining reports, particularly, Progress Report #8 (1971) of the "Special Virus" program of the United States of America. The flowchart is listed as page 61 and on page 2, the "introduction" describes the specific logic of the nature of the flowchart:

"The viral Oncology Area is responsible for planning and conducting the Institute's program of coordinated research on viruses as etiogical agents of cancer. Scientist within this Area not only provide the broad operational management for intramural and collaborative but also conduct comprehensive investigations on specific animal and oncongenic viruses and their interaction with the host cells and apply this information to search for viruses which may be etiologically related to the initiation and continuation of human cancer."

The etiology of AIDS can be found in the "ETIOLOGY WORK AREA" of the National Cancer Institute (NCI). It is reasonable to assume that other malignancies of unknown etiology might also be found in this section of the NCI. Additionally, a thorough review of the records of the National Cancer Advisory Council, the Scientific Directorate, NCI and the Etiology Program Management Group, NCI would definitively reveal the etiology of AIDS and perhaps other diseases.

Progress Report #8 speaks to the prior history of the program. In 1964, the U.S. Congress provided funds to the NCI for an intensified program in virus-leukemia research because many scientists were convinced that an effort to identify viruses or to detect virus expression in human tumorss would contribute to the determination of the etiology of cancer. Progress Report #8 at 2. Using a new planning approach (Convergence Technigue), an overall program aimed at controlling human leukemia and lymphoma was formulated. It is base on the premise that one virus is an indispensable element for the induction (directly or indirectly) of at least one kind of human cancer and that the virus or viral genome persists in the diseased individual. id. The funding level in fiscal year 1971 was $35 million.

Your show has a seasoned practice of correcting error. Congressman Clyburn's office (Chair of the Congressional Black Caucus) has some of this information. Andrea Martin is the person on his staff who most fully understands the ramifications of the discovery of the flowchart, and some of the other critical missing pieces of the laboratory origin of AIDS.

It is my understanding that my presentation in Canada was videotaped. Please contact the Foundation's President, Don Scott (705-670-0180). The flowchart is authenic and the progress reports confirm it. The AIDS virus is a chimera. This point is further underscored when one considers that the genetic sequencing of AIDS contains some of the geneitic sequencing of an animal virus, Visna. Visna is an "Icelandic sheep disease" that was man made by the Germans. How does a recently new Icelandic sheep disease and a monkey disease combine but for human tissue cultures?

Mr. Anderson, the record conclusively reveals the Russians were right. Of course they were right. The record reveals they were in on the development of the preditor virus. This fact is recorded in Progress Report #13 (1976).

The entire purpose of the U.S.Population Policy as outlined in NSSM 200 and executed in NSDM #314 is to stem the birth rate in primarily Africa. Consequently and unfortunately, the progress reports reveal specific experiments that were designed to search for and incorporate "ethnic blood markers" in the preditor virus' s appetite.

As was recently confirmed by author, Robert E. Lee, "Aids: An Explosion of the Biological Timebomb", progress report #8 confirms the "special Virus" program had also isolated 70S RNA and RNA-dependent DNA polymerase (the AIDS workhorse).

It is a further facade the U.S. acted as if it didn't know what was happening in the late 70's and early 80's, when we began to have mystery illness in Manhattan.

Mr. Anderson, you have an opportunity to effect history. Unshackle yourself from the grips of the parallel government and individually challenge the information contained in this letter.

Why does the Archives of the National Cancer Institute only hold three of the fifteen reports of the Special Virus program of the United States of America?

Sincerely,

Boyd E. Graves, JD
Director-AIDS CONCERNS
The Common Cause Medical Research Foundation
216-691-9167 or 216-382-9252

1008 Elbon Rd.
Cleveland Heights, OH 44121-1429

September 15, 1999

Dear Senator Voinovich:

Thank you for your kind letter of August 31, 1999 which I received today. I sincerely appreciate your interest in wanting to "hear my views" on the origin of AIDS.

Over the next several days I will prioritize the overwhelming evidence of the hunt for a preditory virus that selectively kills. Your staff may want to independently verify the existence of the 1971 Flowchart of the "Special Virus". The Flowchart can be located at the National Cancer Institute Archives in Bethesda, MD. Judy Grossberg (the librarian) will copy page 61 of Progress Report #8 (1971) of the "Special Virus" program of the United States of America.

It is my understanding that Phil Park of your staff will direct all further development in this most important matter.

Is the United States Population Policy unconstitutional, or does the state have an inherent right to cull?

The heart and history of depopulation lies in several U.S. documents.

1. The Conference papers from the Ft. Detrick Conference held on April 4- 5, 1969.
2. The "full" testimony of Dr. Donald MacArthur on June 9, and July 1, 1969.
3. The "full" text of the "Special Memo to Congress on Population" by then President, Richard Nixon, July 18, 1969.
4. The "full" text of Henry Kissenger Memorandum #45.

5. The "full" text of the National Security Study Memorandum-200 (NSSM-200).

6. The "full" text of National Security Defense Memorandum #314 (NSDM #314).

7. The fifteen progress reports of the "Special Virus" program.

The progress reports coordinate our experiments, contracts and contractors to the logic of the flowchart. The flowchart is definitive proof of our want to create a leukemia/lymphoma virus of a prototype identical to AIDS. The reasons for the need for AIDS is outlined in the federal documents listed above. It is cruel that the Rockefeller Foundation slants ideology through money and influence to the detriment of the Constitution.

Sincerely,

Boyd E. Graves, JD
Director-AIDS CONCERNS
the Common Cause Medical Research Foundation
cc: Senator DeWine (Michelle Gillcrist)
Rep. Clyburn (Andrea Martin)
Rep. B. Lee (Jennifer Simon)
Rep. Tubbs Jones (Patrick Edmond/ Lance Mason)

September 15, 1999

Professor Peter H. Duesberg, Ph.D.
Department of Molecular & Cell Biology
c/o Stanley/Donner Administrative Services Unit
229 Stanley Hall #3206
University of California at Berkeley
Berkeley, CA 94720-3206

Email: duesberg@uclink4.berkeley.edu
Fax: (510) 643-6455

Dear Dr. Duesberg:

At a recent conference in Canada, Dr. Garth Nicolson confessed his involvement in the "Special Virus" program. He had never before seen the 1971 flowchart of the entire "Special Virus" program.

In light of Dr. Nicolson's overall accomplishments, on behalf of human experiment victims, it may be easy to quickly forgive him and forge a "true" partnership of ordinary people and science.

Dr. Duesberg, your work with retroviruses and the "Special Virus" program is also significant. The AIDS retrovirus is "special" because it has a man made, thus predisposition, for a particular blood marker found primarily in the Black population.

It appears the National Security Study Memorandum- 200 ("NSSM-200") is a "puppet-document" (wish list) of the Rockefeller Foundation to cull humans.

There is great need to assist in the establishment of "real" governments around the world, and the establishment and strenghtening of infrastructures supported by safe and efficient nuclear energy.

The Rockefeller Foundation has throughout this century sought to reduce population with sanctioned support from the United States. The United States has a long history of supporting the depopulation programs of the United Nations and World Bank.One need look no further than George W. Kennan's 1948 top secret State Department Memo.

The lead world unit coordinating depopulation is the United Nation's Agency for International Development ("AID"). The United States has been specifically earmarking funds for AID since 1967. The very same year the National Academy of Sciences ("NAS") admits it began work on an AIDS-like virus. However, the record clearly reveals the "Special Virus" program began in 1962 and an RNA-dependent DNA Polymerease ("RDDP") was isolated as early as 1965. SEE Lee, Robert E; "AIDS: Explosion of the Biological Timebomb?"

Your paper inside the "Special Virus" program of most concern is "RNA: Fact and Fancy". In light of this paper, how could you not know a 'retrovirus' was at the heart of this "contagious mystery illness", during the late 70's and early 80's? It appears your current prognostications about HIV are meant to
mislead in light of your published past.

Many people believe Dr. Robert C. Gallo is an unethical scientist. He may not change unless he is forced. Dr. Nicolson changed when it was HIS family's turn to be culled. Dr. Duesberg we need you. Your recollections of your meetings at the Hershey Medical Center will be invaluable throughout the next millennium.

Your review of the 1971 flowchart is absolutely essential to the restoration of we the people, and the free will of a genuine people under Constitution.

Sincerely,

Boyd E. Graves, JD
Director-AIDS CONCERNS
the Common Cause Foundation

FROM : CRUSADER   FAX NO. : 2166213904   Oct. 05 2000 11:46AM P1

MIKE DeWINE
UNITED STATES SENATOR
OHIO

GEORGE VOINOVICH
UNITED STATES SENATOR
OHIO

United States Senate

WASHINGTON, DC 20510-3504

CASEWORK HOTLINE: (800) 205-OHIO (6446)

November 17, 1999

Boyd Ed Graves
1874 Lampson Avenue, 3rd Floor
Cleveland, Ohio 44112

Dear Mr Graves,

The Offices and Senators DeWine and Voinovich are in receipt of your information regarding the AIDS illness. At this time, neither office intends to pursue an investigation of the matter.

By copy of this letter, I request that you cease contacting staff members though any means other than via the Postal Service. Thank you for your compliance with this request.

Sincerely,

Cole Thomas
Operations Director
Senators DeWine and Voinovich

37 WEST BROAD STREET, ROOM 970
COLUMBUS, OHIO 43215
(614) 469-6774 / FAX: 469-7419

PRINTED ON RECYCLED PAPER

COMMITTEE ON BANKING AND FINANCIAL SERVICES
SUBCOMMITTEES:
CAPITAL MARKETS, SECURITIES AND GOVERNMENT SPONSORED ENTERPRISES
HOUSING AND COMMUNITY DEVELOPMENT

COMMITTEE ON SMALL BUSINESS
SUBCOMMITTEE ON EMPOWERMENT

Stephanie Tubbs Jones
Congress of the United States
11th District, Ohio

WASHINGTON OFFICE
☐ 1516 LONGWORTH HOUSE OFFICE BLDG.
HOUSE OF REPRESENTATIVES
WASHINGTON, DC 20515
(202) 225-7032
FAX: (202) 225-1339

DISTRICT OFFICE
☐ 3645 WARRENSVILLE CENTER ROAD
SUITE 204
SHAKER HEIGHTS, OH 44122
(216) 522-4900
FAX: (216) 522-4908

December 12, 1999

Mr. Boyd Ed Graves
Director-AIDS Concerns
Common Cause Foundation
1008 Elbon Road
Cleveland Heights, Ohio 44121

Dear Mr. Graves:

Thank you for contacting me to express your support for an Congressional Commission to review the origins of HIV/AIDS. I appreciate having the benefit of your views on this matter.

Your letter raised many interesting points. The issue you mentioned in your letter related to concerns that would have a direct impact upon our community. Therefore, I want to personally assure you that I will give serious consideration to this request.

Again, thank you for contacting me regarding this matter. If I can be of assistance in the future please do not hesitate to call upon me.

Sincerely,

Stephanie Tubbs Jones

Stephanie Tubbs Jones
Member of Congress

STJ:pe

Patrick Edmond

PRINTED ON RECYCLED PAPER

U.S. Department of Justice

United States Attorney
Northern District of Ohio

1800 Bank One Center
600 Superior Avenue, East
Cleveland, Ohio 44114-2654

January 20, 2000

Mr. Boyd E. Graves
2700 Washington Street
Cleveland, Ohio 44113

Re: *Boyd E. Graves v. The President of the United States, et al.*, Court of Appeals Case No. 99-4476

Dear Mr. Graves:

Enclosed is a copy of the Appearance of Counsel form which was mailed to the court on January 19, 2000.

Very truly yours,

Lisa Hammond Johnson/pdh

Lisa Hammond Johnson
Assistant U.S. Attorney
216/622-3679

Enclosure

pdh

DENNIS J. KUCINICH
10TH DISTRICT, OHIO

1730 LONGWORTH OFFICE BUILDING
WASHINGTON, D.C. 20515
(202) 225-5871

14400 DETROIT AVENUE
LAKEWOOD, OHIO 44107
(216) 228-8850

Congress of the United States
House of Representatives

Committees:
Government Oversight
Education and the Workforce

www.house.gov/kucinich

February 3, 2000

Mr. Boyd "Ed" Graves
2700 Washington, #1402
Cleveland, OH 44113

Dear Mr. Graves:

Thank you for your recent telephone call. I appreciated hearing from you.

I am sorry to hear about your particular experience with AIDS. It is clear that you have given much thought about AIDS-related issues, and that you are concerned about helping others avoid a similar experience.

I also appreciated your taking the time to share your theories about how governments responded to the AIDS situation. While the theory exists that the spread of AIDS is somehow inspired by some government conspiracy, there is no solid evidence that that is, in fact, true. AIDS affects every race, gender, social class and demographic group in the United States; it knows no boundaries, no limitations, and we ought to focus our energies on finding a medical solution to help those with AIDS and those exposed to the HIV virus.

It is clear to many, however, that the national and international communities were too slow to react to calls for preventive action, and that it is also inherently difficult to education every individual about the consequences of AIDS and the spread of the HIV virus that causes AIDS. Too many Americans have discounted or ignored the very real warnings from governments and health agencies, and our energies should be placed on finding a medical breakthrough to help those in need, such as yourself, and helping to educate the broad general public on the very real consequences of this matter.

You are in my thoughts and prayers.

Sincerely,

Dennis J. Kucinich

Dennis J. Kucinich
Member of Congress

September 8, 2000

Victoria A. Cargill, M.D., M.S.C.E.

Building 2, Room 4B20

Two Center Drive

MSC 0255

Bethesda, MD 20892

Dear Dr. Cargill:

You are the Medical Officer for the Office of AIDS Research for the U. S. Government. You were previously unaware of the evolutionary relationship bewteen AIDS and Visna.

You were previously unaware of a federal virus development program entitled the Special Virus.

After sending your May 15, 2000 letter you chose to avoid communication on this issue. In essence you have personally assisted in furthering the U.S.'s Population Stabilization agenda as is best outlined in P.L. 91-213 (3/16/70), signed by Richard Nixon.

You passed the buck to Dr. Rabson, who has yet to do anything!

It is our position the federal virus program, coordinated by the 1971 flowchart, should receive review from the very highest level of government.

I now understand that because you have been subjected to profanity, you are not going to take any further action. As profane as state-sanctioned murder is, I sincerely believe your actions are more vulgar, than any spoken or written words that seek to respond to this grotesque federal genocide program.

We believe that we are entitled to your best efforts to incorporate the remedies for AIDS (e.g. n-demethyl rifampicin) that are identified in the 15 progress reports of the Special Virus program of the United States of America.

There is nothing more profane than state-sanctioned murder.

Sincerely,

Boyd E. Graves, J.D.

Director-AIDS CONCERNS

the Common Cause Medical Research Foundation

October 14, 2000

Victoria A. Cargill, M.D., M.S.C.E.
Medical Officer
National Institutes of Health
Office of AIDS Research
Building 2, Room 4E20
Two Center Drive
MSC 0255
Bethesda, MD 20892

Dear Dr. Cargill:

As a follow up to our prior conversations and your May 15, 2000 letter, I have been in touch with Dr. Alan Rabson. Dr. Rabson referred me to the Zinder Report to prove the Special Virus program was discontinued in 1977. Dr. Rabson was unaware of Progress Reports #14 (1977) and #15 (1978) which absolutely prove the program continued to produce and proliferate a candidate virus. We can prove this program produced 15, 000 gallons of a "new" immune-suppressing virus.

You state in your letter that you were in posession of the flowchart and some of the Progress Reports of the program.

Please review Phase IV-A of the flowchart and the ensuing decision box. Phase IV-A is designated "Immunological Control" and does appear to hold many of the early secrets with respect to "control" and "reversal" of the Special Virus.

My eight years of research has put me in contact with many distinguished expert medical doctors and scientists from around the world. They are unanimous, the flowchart is the "missing link" in conclusively proving the true laboratory origin of AIDS.

It is my understanding that you are in possession of Progress Report #14 (1977) of the Special Virus program. Please refer to pages 34-47 and the international agreements that are encompassed. There is clear proof the Special Virus was developed in consistent collaboration with other governments, including the Soviet Union. See, Memorandum of Understanding (1972).

There is significant involvement of the National Institutes of Health ("NIH"). Additionally, Dr. Rabson is also a contributor to the program. We do not beleive this federal program should be set aside nor left out of the equation in search for answers and solutions to the AIDS pandemic. We believe there are critical experiments in this program, where, if reviewed, would lead to immediate medical breakthroughs for people living with HIV and AIDS.

We believe you have an obligation to the American people to further review this program and incorporate the wisdom of American and world experts seeking to resolve the AIDS crisis.

In the alternative, we seek emergency funding ($250,000) to conduct a world conference for joint review of the flowchart and 15 progress reports. Please contact me immediately to begin the process for review of this federal program. The delay and orchestrated silence only abet the devastating effects of a "contagious-cancer" virus that selectively kills.

Review of the Special Virus program is paramount in our efforts to understanding the true origin of AIDS.

Sincerely,

Boyd E. Graves, J.D.
Director-AIDS Concerns
the Common Cause Medical Research Foundation
(Don Scott-President, Ontario Canada)

VIA FAX—EMAIL—U.S. POSTAL

December 11, 2000

Victoria A. Cargill, M.D., M.S.C.E.
Medical Officer
Building 2, Room 4e20
Two Center Drive
MSC 0255
Bethesda, MD 20892

Re: Review Special Virus Program

Dear Dr. Cargill:

We are still awaiting a response from you with regard to our request for funding for a conference to review the U.S. Special Virus program.

In May, you referred this matter to Dr. Alan Rabson. Dr Rabson appears to have played a significant role in the development of AIDS and is an inappropriate person for "independent" review of the secret virus program. Dr. Rabson has taken no action for further review of a secret federal virus development program that depletes the immune system. Ending in June 1977, this program produced 15, 000 gallons of a "special" virus. The reports of the program reveal that, at the same time, Dr. Robert Gallo entertained top Soviet biologists interested in "large scale production" of a the new virus. We now know the United States and the USSR had a biological weapons agreement as early as November 1972. The Memorandum of Agreement allowed for the superpowers to concentrate their biological science as a solution to "overpopulation" in Africa. I would be happy to provide you the additional references to the international agreements secretly supporting the culling of the Black Population.

For the record, the flowchart of this program links the research logic of over 20,000 scientific papers.
We are emphatic. We have found the wellspring of the genesis of the AIDS pandemic. It is also important to note your top scientist, Dr. Brian Foley, Director of the Los Alamos AIDS Database, concludes a lack of explanation for the "sudden emergence" of two monkey to man viruses. Dr. Foley, similar to so many, simply refuses to look at the hundred year history of the development of AIDS. Dr. Foley refuses to come to the public podium of debate. We believe the databases of the secret virus program should be interfaced with your databases.

We have seen the realization of eugenics, by any means necessary. In 1945, we had to have the German scientists or the Russians would get them. With the subsequent creation of the Central Intelligence Agency and the World Health Organization, they all began working together to the detriment of people of color.

There were two Manhattan projects. There always has been.
You have already conceded you had no prior knowledge of this federal virus program. However, you then conveniently failed to contact any of our scientists or medical doctors. Dr. Garth Nicolson of the
Institute for Molecular Medicine recently admitted that he was involved in the development of AIDS. You have a duty to the American people to speak with Dr. Nicolson and Dr. Horowitz and Dr. Cantwell. We believe we are entitled to a review of this federal virus program. Scientists and doctors from around the world continue to join our call for review of this secret U.S. program. Colonel Jack Kingston calls the Special Virus program, "the biggest secret in U.S. history." Col. Kingston is chairman of the National Security Advisory Board. The flowchart proves the issue of AIDS bioengineering is a legitimate consideration as to the true origin of HIV/AIDS.

By the way, I note that your former employer (Case Western Reserve University) has entered into an
agreement with Dr. Robert C. Gallo. Dr. Gallo refuses to explain his role as a "Project Officer" for the U.S. Special Virus Program. Dr. Gallo surgically excludes his role in the development of AIDS in his autobiography. Dr. Len Horowitz suggests and the progress reports of the program confirm, Dr. Robert Gallo is the center of the Hegelian dialectic.

The program's flowchart is the quintessential missing link in establishing conclusively the laboratory
origin of AIDS. Proving the true origin of AIDS is necessary for the immediate medical breakthroughs it might provide for people living with HIV/AIDS and other illnesses.

We again seek your immediate attention to our call for review of this secret federal virus development program. The decisions reached by the secret program in conjunction with Phase IV-A ("Immunological Control") of the flowchart appear to hold the most promise for better therapy and treatment.

Please contact me to further discuss the relevant issues identified in our continued correspondence.
Thank you for your interest in the welfare of humanity and our collective interest in deactivating AIDS.

Sincerely,

Boyd E. Graves, J.D.
Director-AIDS CONCERNS
the Common Cause
Medical Research Foundation
1-888-842-6419

enc: Cargill, V.A. May 15, 2000 letter

cc: Dr. David Satcher, U.S. Surgeon General

Dr. Eric Goosby; AIDS Director U.S. Surgeon General

Speaker Dennis Hastert

Senator Trent Lott

June 18, 2000

Dr. Brian Foley

AIDS Database Supervisor

Los Alamos National Laboratories

Los Alamos, NM

Dear Brian:

WHO is Tanmoy Bhattachary? He is unaware of the evolutionary relationship between AIDS and VISNA. The glycoprotein of AIDS is identical to VISNA's.

There is no mention of lentiviruses in your database. There is no mention of mycoplasma in your database. There is no mention of VISNA in your database.

Please review the following citation: Proccedings, National Academy of Sciences, Vol. 92, 3283 - 3287, (April11, 1995).

AIDS has evolved from a 'man-made' "prototype" virus. Your database makes no reference to the 1971 AIDS flowchart of the "Special Virus" program.

How can Los Alomos exclude all of the relevant research of this 'secret' virus program?

Please contact me via phone at 216-561-1967. I am gravely dismayed that Mr. Bhattachary sent the HIV database in a clear cover!

There are still 'missing secrets' at Los Alomos!

Sincerely,

Boyd Ed Graves, JD

Director-AIDS CONCERNS

the Common Cause Medical Research Foundation

December 17, 2000

Dr. Alan Rabson
Special Virus Program
NCI

Dear Dr. Rabson:

Since May, you have made yourself "unavailable" in providing a response to the flowchart of the secret federal virus development program. Judy Grossberg of your archives will send over to you a copy of the cover of the 1978 report. The 1978 report reveals a "vibrant" virus program contrary to your assertion the program was "suddenly terminated" in 1977.

Please let me know when I might be able to deliver the flowchart to you on this coming Thursday, December 21. I will be delivering the flowchart to several other person in the NIH/NCI complex and would sincerely appreciate 20 minutes to further highlight the thoroughness of our research into the "Special" virus. We need your support of our effort to host a conference of the growing number of experts around the world who have reviewed the flowchart and one or more of the progress reports of the program. On Friday, Dr. Cargill alerted us that Director Linda Jackson would be contacting us this week in our emergency request for funding. Our core of experts is led by Dr. Len Horowitz , Dr. Alan Cantwell, Dr. Robert E. Lee, Dr. Dean Loren, Dr. Vincent Gammill and Don Scott, President of the Common Cause Medical Research Foundation.

The August 1999 unveiling of the flowchart to the international medical and scientific communities was one of the greatest document presentations in history. The flowchart provides the key to understanding the program's delineation of experiments for "specific purposes".

The flowchart is the "research logic" of a secret Manhattan project that spent $550 million to make AIDS and other illnesses. In actuality we believe the program spent even more.

Dr. Rabson we accept the premise that you may speak for the NCI and NIH on this issue. We believe you are fair enough to accept the flowchart from me on Thursday.

Thank you for your continual commitment to the welfare of the human race.

Sincerely,

Boyd E. Graves, J.D.
Director-AIDS CONCERNS
the Common Cause Medical Research Foundation
1-888-842-6419

cc: Dr. Cargill

Dr. Goosby

Dr. Horowitz
Dr. Cantwell

Dr. Lee

Dr. Nicolson

Dr. Loren
Don Scott

November 10, 2000

Steve Koff

Cleveland Plaind Dealer

Washington Bureau Cheif

Dear Mr. Koff:

Enclosed please find an email from Professor Robert E. Lee. Professor Lee can be reached at 309-797-6027. Professor Francis Boyle, author of the U.S. Anti-Biological Terrorist Attack ACt of 1989 gave an interview to your federal court reporter.

Your newspaper continues to set aside public notification of a "secret" virus program because you have been convinced the program did not exist.

Even after providing you with the flowchart of the program you still want your readers to believe that "uncontrollable African lust" led to the pandemic. If that is true, you should be able to demonstrate to your readers that the 15,000 gallons of AIDS, made by the U.S., are still on a shelf at Fort Detrick.

The American people are entitled by the Constitution to a "neutral" press. Just because you do not yet understand the flowchart, you can not deny it exists.

After you speak with Professor Lee, please let me know, I will then provide you with Professor Boyle's number.

Dr Alan Cantwell, Dr. Len Horowitz, Dr. Garth Nicolson and Dr. Alan Rabson are all in the wing to assist this process of further exposure of a federal virus program.

Why have you yet to contact Col. Jack Kingston, Chairman of the National Security Advisory Panel? According to Col. Kingston, the Special Virus program is the "biggest secret in U.S. history".

Your National Editor, Flora Rathburn, does not believe the CIA conducts propaganda. In this regard, you have accepted hook, line and sinker an 'out of Africa' origin of AIDS. Whether you believe me or not is irrelevant. However, when you are provided credible, supportive information from the experts, you have a Constitutional obligation to report to the American people.

We have found a flowchart and progress reports of a secret, Manhattan-style virus program of the federal government. The American people need your journalistic inquisitiveness for the continual record of truth and fact. It is time to 'pay attention' to the men behind the curtain of AIDS. Click your heels and start doing your job for the good of the welfare of humanity.

Please give me a call after you have conferred with Professor Lee. Thanks
Steve.

Sincerely,

Ed Graves
3844 E. 140th Street
Cleveland, OH 44128
216-561-1967

ps. I understand that you have a copy of the program's flowchart.

November 24, 2000

> Dr. Graves,
> I can't say that I agree or disagree with your argument that AIDS was created in a laboratory with the specific purpose of destroying our race. At this point I don't really think it matters where it came from. The fact is that it's here and we have to fight it. I'm more focused on eliminating it than revealing its origin. What are your feelings on HIV prevention in the African American community? How can we make sure that no more African Americans become infected?
> Lois Jones <lgj@po.cwru.edu>

Dear Ms. Jones:

Thank you for contacting me regarding World War AIDS. I note from your email that you appear to be affiliated with Case Western Reserve University. I have been in touch with Dr. Victoria Cargill of NIH formerly of your university and Dr. Michael Lederman.

Your email raises a degree of scepticism that appears to demean the eight years of research that I have completed on this issue. In short, I would appreciate you contacting me by phone. My number is 216-561-1967.

My position is that the "true origin" of AIDS IS relevant to our current hunt for cures and deactivation. I would be happy to provide you a copy of the flowchart of the federal program today! The AIDS virus is the "candidate virus" sought by the United States in a secret virus development program entitled the "Special Virus". In indentifying the "true origin" of AIDS, we have our best avenue to medical breakthroughs which may eventually deactivate the virus. Thus, your goal of zero new infections in the African American community would be achieved. I solicit your support in joining our efforts calling for "independent review" of this federal program. The government of South Africa is the latest superpower to join this call.

According to the flowchart, the United States has maintained "inhibitors and reversals" of AIDS for the last thirty years. The knowledge gleaned from the progress reports of the program would immediately lead to medical breakthroughs for those those of us living with the virus. I strongly support our local programs of condom distribution and safe sex AIDS education and have personally imple mented a sex abstinence scenario. However, I do feel that the fastest way to defeat AIDS for all people (not just African Americans), is through review of the federal virus program that began officially in 1962.

For the record, the federal virus produced 15,000 gallons in 1977. Then, and only then, did we have "dualing epidemics" of AIDS in Africa and Manhattan. It is clear, the keys to unlocking the mystery illness, lie within the 15 yearly progress reports of this secret program, and the 20,000 scientific papers that are linked through the "research logic" contained in the program's flowchart.

Our search for methods to ensure no new infections should not be in the jungles of Africa or the bath houses of Manhattan, but in the laboratories of collaborating institutions of this federal program. In this regard, Dr. E. Frederick Wheelock conducted many experiments on the blood markers of African Americans at your institution.

If you are truly interested in no new infections, please give me a call. Dr. Michael Mederman refuses to acknowledge the existence of the federal program, even after Dr. Victoria Cargill located the flowchart and progress reports in the archives of the National Cancer Institute!

We can clearly show the AIDS virus has been "bioengineered". Now we must work together to take it apart. In my opinion, THIS is our best route to deactivating AIDS for all Americans. Thank you for contacting me and I eagerly look forward to your call.

Sincerely,

Boyd E. Graves, J.D.
Director-AIDS Concerns
the Common Cause Medical Research Foundation (Ontario, Canada)

cc: Victoria Cargill, M.D.
Michael Lederman, M.D.

July 24, 2000

Jonathan D. Moreno, Ph.D.
Center for Biomedical Ethics
University of Virginia
Box 348 Health Sciences Center
Charlottesville, Va. 22908

Dear Dr. Moreno:

At the heart of the HIV genome is a mycoplasma. It is identical to the mycoplasma located at the heart of VISNA. The ongoing exchange of citations with Los Alamos is awesome. This week Congressman Conyers (aide: Joel Seigel) is going to comment on the flowchart and the secret virus development program, the "Special Virus".

All of my scientific and medical citations are coordinated back to the creation of the U.S. "Laboratory of Hygiene" in 1887. Mycoplasma (PPLO) was isolated in 1889. I would like to present the flowchart and progress reports in England in September when the royal society is hosting a world conference on the origin of AIDS. Last August I unveiled the flowchart at a medical research conference in Canada. Dr. Victoria Cargill (NIH) has INDEPENDENTLY located the flowchart. Dr. Brian Thomas Foley (Los Alamos)and I are engaged in an intense exchange of information. I need to be de-briefed over the 8 years of irrefutible research.

The U.S. authorizes 'population stabilization' on March 16, 1970. (p.l. 91-213). It is part of a follow up to Nixon's 'secret memo to Congress' (July 18, 1969). According to the U.S. patents, AIDS is not a virus, it is a mycoplasma. See, U.S. patent #: 5242820.

Prior to any conversation, please review: Proceedings, National Academy of Sciences, Vol. 83 pp. 4008 - 4011 (1986). It is easier for me to speak via phone than email. Please contact me as soon as possible. I have been able to make copies of most of my materials. Thank you for your interest in humanity.

Sincerely,

Boyd Ed Graves, J.D.

Director-AIDS CONCERNS
the Common Cause Medical Research Foundation
Don Scott, President 1-705-670-0180

VIA EMAIL

From: Norton Zinder <zinder@mail.rockefeller.edu>
To: Boyd E. Graves <ed@boydgraves.com>
Date: Monday, November 06, 2000 10:50 AM
Subject: Fw: the Zinder Report '77

Since the report is more than 20 years old, I would like to know who you are and what you want it for.

It certainly is dated, The only report I know of is the 1974 Zinder report on the NCI Virus Cancer Program.

Norton D. Zinder

—

Norton D. Zinder
Rockefeller University
1230 York Ave
NY, NY 10021
212 327 8644 Ph
212 327 7850 F

VIA EMAIL

From: Boyd Graves <boyded@xcelnet.net>
To: Dr. Norton Zinder <zinder@mail.rockefeller.edu>
Date: Monday, November 06, 2000 10:50 AM
Subject: Fw: the Zinder Report '77

Dear Dr. Zinder:

Alan Rabson suggested that if I were to read your report I too would be convinced the "Manhattan-style project, the "Special Virus" program was suddenly terminated in 1977, and could not have played any role in the etiology of AIDS.

It is our position that progress reports #14 and 15 (1978) thoroughly prove the continuation of the program. It is reasonable to assume in light of the15 progress reports of this federal program, the program proceeded on to complement vaccines. In the very least we believe we are entitled to accountability with respect to the 15,000 gallons of AIDS this program produced between March 1976 and June 1977.

My name is Boyd E. Graves. Last year I found the program's flowchart. It coordinates over 20,000 scientific papers in the "creation, "production" and "proliferation" of an immune virus. Last August, in my position as Director-AIDS CONCERNS, for the international medical research foundation, Common Cause, I unveiled the flowchart.

I would be very happy to provide you a copy of the flowchart. I would also like to provide you with at least the first 60 pages that precede the flowchart. Every experiment, including those of Dr. Gallo and others, is linked to each of the phases of the flowchart!

Colonel Jack Kingston, Chairman of the National Security Advisory Board, believes this program is the "biggest secret" in United States history. We believe this flowchart is the definitive proof of a "Wizard of Oz-curtain" surrounding the 'black budget' program that eventually complemented vaccines(simultaneously).

It is our position that the "Special Virus" (AIDS) was designed with Congressional Appropriations throughout its development. The necessity to cull Africa is outlined by Richard M. Nixon on July 18, 1969. However, it is March 16, 1970 and his signing of PL91-213 that now appears to have been a
victory party for eugenics.

It is our position the databases of the "Special Virus" program should be interfaced with the supercomputers of the AIDS DATABASE in Los Alomos, NM.

The NCI archives only houses 3 of the 15 reports! We have six! We believe all 15 should be immediately reviewed! We have within our reach the ability to "deactivate" AIDS before the millenium. We can clearly show the program continued in accordance with the research logic as outlined by the flowchart.

I would be happy to provide you aditional information. I am the third son of a coal miner and a school teacher and I was raised in Northeast Ohio. In 1971 I accepted an appointment to the United States Naval Academy and am currently celebrating my 25th anniversary. Just prior to finishing my last semester of law school, I learned I had AIDS.

If they didn't want me to seek the truths, about AIDS, then they shouldn't have given it to me. Dr. Rabson has an obligation to Dr. Victoria Cargill, NIH, Office of AIDS Research. On May 15, 2000, Dr. Cargill indepedently located the flowchart through the NCI archives. She had been previously unaware of the program and turned the materials over to Dr. Rabson. Dr. Rabson is purposefully stonewalling because he is listed in the progress reports.

We believe Dr. Cargill has an obligation to 'want to know more' upon locating the flowchart and progress reports as she reported in May. We believe the flowchart is the "quintessential" missing link in connecting
the
direct evidence of the scheme and its execution in massive premature death, precipitated by the state.

I will send you the flowchart or you can get it from Dr. Cargill. I will also send you the first 60 pages that precede the flowchart. We CAN deactivate AIDS.

Sincerely,

Boyd E. Graves, J.D.
Director-AIDS CONCERNS
the Common Cause Medical Research Foundation
Don Scott, President 1-705-670-0180

VIA EMAIL

August 31, 2000

RE: REVIEW SPECIAL VIRUS SOLVE AIDS ORIGIN

Dear Professor Lee:

On page 283 of Progress report #8 (1971) of the Special Virus program, the United States confirms the special virus program was inoculating monkeys with Reovirus 3, in a mixture with Burkitt's Lymphoma. According to the progress report, **the 'primary objective was the investigation of the possible oncogenicity of selected human prototype viruses in primates in conjunction with the use of co-carcinogens'.**

We believe this is further 'direct evidence' of the development of Mammal AIDS. The program's flowchart reflects "progressive steps" in the development of a contagious cancer that selectively kills. The first sixty pages of the report are a narrative overview of the flowchart.

Instead of doing the noble thing and seeking a review of the program, Dr. Rabson of NIH simply will not reply. This allows Dr. Cargill the 'out' of saying that she has done something, while P.L. 91-213 (3/16/70) allows for the further extermination of the Black population. Proof of the 'racial design' of AIDS (according to the scientific literature) lies in its affinity for the OKT4 epitope deficiency in the Black population.

The research trail remarkably shows continuous development of a selective contagious cancer (mycoplasma) going back to the 1887 U.S. mandate for the creation of a **Laboratory of Hygiene**.

Negroes got their freedom from slavery, however, 'we got something else for their sex crazed, poverty baby-making libido'.

It is written, the hunt for the AIDS virus was driven by the insecurity of deep-seeded racism, based on sexual inferiority. In other words, since they (Blacks) like to copulate so much, and they are so good at it, we will devise a scheme (George W. McKennan, 1948) to assist the Negro in copulating himself into extinction! Negroes may be king in the bedroom, but we's the king with the petri dishes.

Just ask Craig Venter and his 300 gene mycoplasma trap that will soon be set, now that he definitively knows they have decreased the probability of "blowback" onto White peoplpe to a percentage of non existent. The elimination of the Black population is well under way. For them, it has always been about racial color and sex.

WITHOUT THE FLOWCHART, THEY WOULD WIN. For the record, every experiment and every contract of the special virus program is coordinated through the flowchart. I will be happy to show you where this is in PR#8.

Now that Dr. Hayflick cannot deny the mycoplasma computer, the United States can not say the small pox vaccine and the Manhattan hepatitis B vaccine were "accidentally contaminated" with the mycoplasma, virus-like particles. We believe the special virus computers should be interfaced with those of Los Alamos AIDS Database.

For the record, HIV was originally called "Leukemia/lymphoma" Virus.

HIV (leukemia/lymphoma) virus is a single molecule with an HIV part and a mycoplasma part? Sonigo's (Cell 42, 369 - 382 (1985)) note added in proof says it all! Dr. Cargill, NIH's Medical Officer for the Office of AIDS Research was unaware of the evolutionary relationship between the Icelandic sheep disease, Visna (sheep AIDS) and Human AIDS. It is strange our scientists and medical doctors did not immediately look for prior explosions of animal models of "wasting", even though Visna had been isolated and characterized since 1949!!

Now she is too embarrased to take affirmative steps to see why the progress report was recommending n-demethyl rifampicin to inhibit AIDS. I don't get it, a medical officer unwilling to review relevant, credible information about potential breakthroughs in the greatest genocide, state-sanctioned program in the history of the world. Why? Probably because not enough (of us (Blacks)) have died The NIH/NCI recalcitrance to investigate legitimate remedies for the pandemic have Adolph H. smiling from below.

Sincerely,

Boyd E. Graves, J.D.,

Director-AIDS CONCERNS, the Common Cause Medical Research Foundation

VIA EMAIL
August 31, 2000

Subject: Re: REVIEW SPECIAL VIRUS-SOLVE AIDS ORIGIN

Dear Dr. Alan Cantwell:

In reference to your mycoplasma mention in 1988, I have just reviewed that chapter (1988) and do not find it. However, endnote #42, Beatrice Hahn and Bob Gallo paper.

endnote #57, highly immunogenic p66/p51 as the reverse transcriptase of HTLVIII/LAV

endnote #58, expression and processing of the AIDS virus reverse transcriptase in E Coli.

endnote #45, AIDS retrovirus induced cytopathology: Giant Cell Formation...

endnote #49, B.HAhn, locationand chemicla synthesis of a binding site for HIV-!

endnote #52, Synthetic CD4 peptide derivates that inhibit HIV infection and cytopathicity (WHERE DO I GET SOME)?

endnote #59 WOW, I will try to get this one tomorrow, I may not sleep tonight.

endnote #88 David Ho, Pathogenesis of infection with HIV

and on and on and on....

EIA transcription induced: "Enhanced binding, rectal transmission in cows and sheep (goddamn buttfuckers)

two elements in the bovine leukemia virus long terminal repeat that regulate gene expression (#26), based on the 1898 paper, they will be mycoplasmas. I definitely have to find this paper tomorrow.

FOR THE RECORD, THE FIRST DECISION POINT OF THE FLOWCHART IS TO FIND A "VIRUS PARTICLE" (IN THE HUNT FOR A "CANDIDATE" VIRION VIRUS FOR LARGE SCALE PRODUCTION). Have you not spoken with Scully about the 60,000 liters that were produced?

Perhaps you should maybe read his book, Queer Blood, pg. 45.

what is adenovirus? Can't we show recombination here? What is the RIP (Recombination Identification Program).?

10,000 students reached, no converts? Are you aware that Don Scott is on the staff of John Martin's Institute? Are you aware that Don Scott is on the staff of Garth Nicolson's Institute? So is the World Wildlife Fund!!

Mulder, it appears that the 1898 isolation of PPLO's was quickly followed by the 1904 appearance of acute lymphocytic leukemia (ALL)—All juvenile Blacks and Browns. This is why you get a distinction with HTLV-I being a "mature" Blacks and Browns (Caribbean and Japanese) disease.

What's up with this Oxygen Therapy Blass (OTB) which originated in Germany in 1898 at the **"Institut fur Sauerstoff Heilvahfahren".**

Where is your research on the U.S.'s 1887 mandate for a "Laboratory of Hygiene"?

Let's get serious. Binley, J. (1997) HIV-Cell Fusion. The viral Mousetrap. Nature 387: 346 -348. Craig Venter is right we can kill them with even less genes than we thought. This 300 gene mycoplasma is hideous. Level the field and provide comment on the citations. Gallo's 1971 "Reverse Transcriptase" find in human cells does not appear in his book either. It does appear that the history of "virion" (1964) predates Gamato.

See also, Turner, A.W. **(1935)** A study on the morphology and life cycle of the organism of PPLO, J. Bacteriol. 41: 1 - 32.

(Magnus Kunta) Boyd E. Graves, J.D.

p.s. While I was out herding my feline friends, I found in the first sixty pages of Progress Report #8 (1971) why they have AIDS.

VIA EMAIL

September 02, 2000

Subject: Re: REVIEW SPECIAL VIRUS-SOLVE AIDS ORIGIN

Dear Dr. Bob Lee,

Where is the conclusive proof of SIVcpzCON being the parent of HTLV-III? Since Visna was first created in the 1930's and HIV shares the "identical central region"; with Foley's conclusion that SIV transformed into HIV in 1931, it simply defies common sense logic that an Icelandic sheep and an African monkey comingled on their own!!

(Now I see that Brucellosis (Sir David Bruce) is also from 1930.)

Where is your comment on pages 104 - 106 of PR#8 (1971)?

Where is your comment about Dr. Robert Gallo's 1971 paper, "Reverse Transcriptase in Type-C Virus Particles of Human Origin", page 335, #650?

Where is your review of the first 60 pages leading to the flowchart that requires a **"virus particle"** base for large scale production? See, Flowchart (First decision point).

Where is your review of U.S. Patent No.: 5242820?

Professor, all this virology is very good, however I feel it is mis-directed. Please post copies of any letters that you have sent to Congress, or any letters that any of the 10,000 people you have informed about the lab origin of AIDS have written.

THE WORLD IS STILL AWAITING YOUR REVIEW OF SONIGO'S 1985 PAPER. According to Vincent, you now have it in your possession.

Your silence on this paper is troubling. If I understand your latest virology, Dr. Foley's phylogenetic trees should have a co-relationship with minks, yes or no.

If the patents are so important, when will you hold class?

By the way, I am used to being ignored. We will get the answers that we seek in the courtroom.

You have suggested that if I am so sincere, I should sue the United States for genocide in the World Court. In May of 1985, the United States Senate passed a resolution that allows the United States to exempt itself from jurisdiction over the Genocide Treaty. The Treaty was signed in March of 1986.

Have you ever reviewed Senator Jesse Helm's paper, "Strategic Importance of Africa" written for the Selous Foundation?

I feel we need more help with the social science of AIDS?

Sincerely,

Boyd E. Graves, J.D.

ps. n-demethyl rifampicin, page 24 PROGRESS REPORT #8 OF THE SPECIAL CANCER PROGRAM (1971)

visna, pages 39 - 40 1904, The 'Station for Experimental Evolution' founded by Charles and Gertrude Davenport

Special Biological properties do not come from nature or spontaneous regeneration. (M.K.)

Where does the Chronic Fatigue human lentivirus fit on the tree?

VIA EMAIL

September 03, 2000
Subject: Re: REVIEW SPECIAL VIRUS-SOLVE AIDS ORIGIN

Dr. Cantwell:

This is crazy! If what you say is true, and it is not, then the govt will have all the 60,000 liters available. It is our contention that the special virus was attached as complement to the small pox vaccine and the hepatitis B vaccine (They can not argue mycoplasma 'contamination'). Africa and Manhattan show evidence of mass infection following mass vaccination. As smallpox ended, AIDS began. As gays were recruited and vaccinated, AIDS began.

All of the progress reports show relevant discoveries of a continuous program! You even suggest that the Manhattan portion of this program proves the 1973 Asilomar conf. sought and succeeded in creating old cancers in a new virus. All of your published literature supports the existence of a federal virus development program! Your work has extensive support in Great Britian and elsewhere.

Your divergence supports the proposition that you have a role in continuing the quagmire. . . . Simply put, THE FLOWCHART DESERVES FURTHER SCRUTINY!!!!!!!!!

It is a very sad day in the world that you are now working inopposite to affecting "neutral" review of this document. Somewhere along the line you have chosen to dismiss a century long progressive hunt for a contagious mycoplasma cancer that selectively kills.

Now that white fags aren't dying, you are content that it is ok to get rid of Africans/Blacks? You list in your book all of your friends who have died from AIDS. How many of them are Black?

Your flip flop means that Black people will have to go it on their own. Your prior reference to yourself as a 'gay nazi' must now be reconsidered. Our collective inability to call a spade a spade allows for the further elimination of the Black race. I recognize that this is not a priority issue for someone who is not Black nor infested with AIDS.

So that I have a final understanding of your position, do YOU support our call for review of this program? Progress Report #15 (1978) proves the program was not terminated in 1976!!!!!!!!!

You never once submitted a review of Nixon's 3/16/70 proclamation of a "population commission". You have never reviewed Nixon's hideous July 18, 1969 "secret message on population". I note that you did not correct your assessment that the mycoplasma lab began in 1969 (p.256, PR#8). It began in 1964. Without that tidbit, others were previously saying that mycoplasma could not be detected until the perfection of PCR in 1986!!!

Some of us BELIEVE you in your audiotape with Garth Nicolson, "Virus Makers of the CIA". I remember VERY vividly last year your absolute astonishment to hear that Garth was at the Canadian Conference. "What was HE doing there?"

It is a very, very sad night for humanity. You will fight to be subpoenaed for a deposition, there is no doubt. After twenty years of research into the origin of AIDS, you now conclude that you are unsure! In other words it is plausible that after thousands of years, Africans only recently became susceptible to a monkey/Icelandic sheep mycoplasma.

Your silence, through all your published literature, on the 1887 mandate creating a U.S. Laboratory of Hygiene, in actuality speaks volumes. I have known all my life that White people will never seek to support a position offered by a Black man. The necessity for the AIDS Bioengineering lawsuit is crystal clear.

Sincerely,

Boyd E. Graves, J.D.

Lead plaintiff AIDS Apology

VIA EMAIL

September 4, 2000

RE: REVIEW SPECIAL VIRUS, SOLVE AIDS ORIGIN

Dear Dr. Robert Lee :

When you compare mycoplasma and visna and AIDS, you will find the same thing! For the record, I am stating that this federal virus program was seeking a "candidate" contagious cancer susceptible to people of color. You never reviewed, The NAS Proceedings with respect to a "rearranged" cancer gene in gay Kaposi Sarcoma. Nor any of the supportive evidence of the evolutionary relationship between the German "visna" (WASTING) and human wasting.

It was relevant when gays were being 'wasted', it does not appear to be relevant now to continue to review the irrefutible, direct evidence of the handprint of man in the African genocide.The search for and maintenance of a "contagious cancer" has made many people rich and famous. No one appears to be interested in bringing this matter to an end.

I will assume that you have read the first sixty pages that precede the flowchart in Progress Report #8 (1971) of the Special Virus program. Recently, you made an inference that 'only Ed' believes the n-demethyl rifampicin remedy for AIDS. Again, I will direct your attention to page 24 (PR#8).

LOOK what they say about VISNA. Your hesitance to post a review of the Sonigo paper speaks volumes!!!!!!! Don't forget, the supercomputers at Los Alamos could not get around this paper. This is the real reason Dr. Brian Thomas Foley of the GOVERNMENT stopped responding after a 10 month dialogue!!

There is no middle ground. They did it, the flowchart conclusively proves it. As much of a reviewer as you and Mulder are, I will await your critical review of the document. START by getting around the necessity of the first decision point of the "RESEARCH LOGIC".

Adolph is truly smiling from below! BTW, whatever became of HeLa virus? In the master world of eugenics, I and all the rest of THEM, are scourges in need of a blight.

I think by now you know I will not go quietly. I do not look forward to the personal degradation, but Black people are entitled to a voice in this discussion, particularly those of us who are gay, particularly those of us who have AIDS.

The Special Virus program is the greatest assault on humanity in the history of the world. How hollow the war cry of 'never again'. I will continue to fight the culling of the Black race. This is personal.

Sincerely,

Boyd E. Graves, J.D.

Lead Pla intiff for Global AIDS Apology

September 4, 2000

Dear Vincent(Gammil):

Just as i suspected, you ARE our only hope. I will share this true response from Mulder (Bob Lee) with POZ magazine tomorrow and the BBC. Now we are getting somewhere. Confuse, dilute, befuddle, anything to keep Black people dying. All of this is a big joke to those who are not Black and do not have AIDS.

Mulder has shared his research with 10,000 people, and not a single letter of inquisition. His answers below are in the correct context of someone who is making a mockery of Black people. His prior statement that the flowchart is not relevant is equal to the tilted, racist mentality that allows pompous Whites to dismiss Blacks by definition. I will relook in his book for his quote on the 'identical central regions' of AIDS and VISNA IT is the same conclusion reached by Gallo and Sonigo. See (Gallo, et al , January 11, 1985, Science and Sonigo, et al Cell 42, 369 -382 (1985) and the 1995 Proceedings of the National Academy of Sciences , Vol 92, pp. 3283 - 3287. (HIV has identical genetic sequencing as to the "prototype" 1931 strain of Visna (strain 1514).

We are not entitled to any legitimate review of anything that does not serve to keep the water muddy. Just watch.

Soon others of you will join the chorus that the federal virus development program is not relevant. We all know Africans need more intimacy than that which is provided by their seven wives, but apparently one needed a little more.

If HIV evolved from SIV, where is his proof?

LOOK at the "NOTE ADDED IN PROOF" in the Sonigo paper, page 382. It is the motherland of mycoplasma consistent with Fort Detrick's April 4 - 5, 1969 conference. On this point Horowitz is brilliant.

I have seen this type of mentality from White people all my life, they always take their ball and run home. If you don't see it their way, they won't play with you anymore. Screw Ed, let him herd pussy (a living death! :) Please pass on to Mulder that it is theoretically impossible to herd pussy because they have all been gathered up by the Special Virus program, See page 44, Progress Report #8 (1971) of the Special Virus program of the United States of America. Therefore, the FIV and HIV patents would be related. ARE THEY? (for the answer boys and girls, go to Gallo's patent on HIV US04647773). Once Mulder and Scully found that mycoplasmas regulate the repeat sequences of HIV, it has been a different ballgame!

However, none of this matters because only scully and mulder hand out pussy herding certificates, and Ed Graves is not going to get one! (nor did I want one!). Mulder's answers as listed seek to further discredit him and scully. As I suspected, this is all by design. Screw Black people and especially that Ed Graves. (What an asshole I am to seek review and further accountability of a "linking" flowchart and progress reports of a secret mycoplasma program).

Again as a researcher I am searching for the earliest paper on Visna. What is the earliest documented scientific paper on Visna. There is but one definitive paper, I can't find it. I understand that Visna erupted in the sheep population of Iceland in the 1930's. Thus, it is logical to now sweep the PUBMED's of Iceland, and translate the early paper and subsequent ones. Where will those references lead?

We have brought them the flowchart and developing scientific discussion of a federal virus development program. To Alan and Bob, I am very sorry if I have offended you in an unprofessional manner. Please understand aggressiveness and activism from the standpoint of a social introvert (reject) with AIDS. I recognize my personal struggle in fighting the AIDS mycoplasma is because I have it. I perhaps would not be in this arena otherwise. However, our best united efforts will bring about a better life for me. I am very sorry if I appear bombastic, brash and arrogant. I really don't know any other way to deal with seeking closure on this issue.

This is a call to regroup and reenergize our efforts in a fashion that wil continue to answer questions for the good of the good people of every genetic code.

"The answers will come from no one discipline alone". **PROGRESS REPORT OFFICE OF THE ASSOCIATE SCIENTIFIC DIRECTOR FOR VIRAL ONCOLOGY (OASDVD), 1971, J.B. Moloney, Ph.D., p.45.**

October 23, 2000

David M. Hillis
Institute for Cellular and Molecular Biology
School of Biological Sciences
University of Texas
Austin TX 78712

Dear Dr. Hillis:

We sincerely believe we are entitled to a response to our July 29, 2000 email. We sincerely believe we have raised significant challenge to your "Origins of HIV" article, Science, Vol. 288, 9 June 2000. In short, it is our position that your failure to review the flowchart and progress reports of the U.S. "Special Virus" program is glaring proof of your acceptance of the secret virus development program (1962 - 1980).

We again challenge you to public discussion of the federal virus program and the century long development of mycoplasmas (and Visna). On page 24 of progress report #8 (1971) of the "Special Virus" program, "n-demethyl rifampicin" was identified as being able to stop the federal virus.

We sincerely believe an independent global review of the flowchart and progress report of the federal virus program will lead to "immediate medical breakthroughs" for people living with HIV and AIDS (and other illnesses)!

Because of e.g., Rockefeller, Kissinger and Nixon (PL91-213, 3/16/70), every hour 200+ Black people are being killed.

There has never been a greater, purposeful assault on humanity than that which you support by your orchestrated silence, and "eugenics science" based on a flawed phylogenetic tree with a slipknot! We sincerely believe the AIDS databases at Los Alamos should be integrated with the databases of this federal virus program.

Soon we will not allow you to ever again, further 'goose-step' on the destiny of the Black race. Please contact me for a public airing and discussion of our 'direct evidence' of the laboratory origin of AIDS.

We demand that our evidence be included in a joint effort to deactivate AIDS. If 200+ Caucasians were being killed by this nation state every hour, I am sure you would not be sitting by attempting to provide "voodoo statistics" as you currently are undertaking.

I look forward to your immediate response in the interest of the welfare of all humanity.

Sincerely,

Boyd E. Graves, J.D.
Director-AIDS CONCERNS
the Common Cause Medical Research Foundation
Don Scott, President, 705-670-0180

1-888-842-6419 <boyd@boydgraves.com> (Cleveland, OH)

August 27, 2000

Dr. Len Hayflick/ U.S. Mycoplasma Laboratory

Stanford University

Dear Dr. Hayflick:

Thank you for providing the citation to the 'Hayflick Medium". I will view the paper tomorrow. I suspect that it is similar to the PPLO artificial medium you spoke of in the Proceedings (NAS) Vol 48, pp. 41-49, 1962. In 1972, in Progress Report #9 of the Special Virus program, you reported that one of the major findings of the mycoplasma laboratory was;

(2) All tumor culture data, patent history, and cell storage information is now stored in a computer. (p. 287).

The flowchart I mentioned is the leaflet, blueprint of the "research logic" of the 15 progress reports of the program. It has been reviewed by some of the top medical doctors and scientist in the world and many will state it is the research logic of a federal virus development program (See, e.g. Horowitz, or Cantwell or Lee). In 1977, according to author/expert Dr. Alan Cantwell, quoting Richard Hatch, this program concluded the production of 60,000 liters of the special virus.

We sincerely believe an immediate review of this program is warranted. I believe pursuant to the unchallenged scientific paper of FEMS Microbiol Lett 128, 63 68 (1995), the special virus is a mycoplasma program that regulates the repeat sequences of AIDS.

Additionally, it appears that AIDS has an evolutionary relationship with an **"ungulate wasting disease of Iceland"** called **Visna**. See Cell, Vol. 42, 369-382, August 11, 1985. The Los Alamos supercomputers cann ot dispel this critical scientific paper, which is supported by the Proceedings (NAS) of the United States of America. See Proc NAS, Vol 92, 3283 -3287, (1995). Your colleague, Dr. Howard Clark has also recently been in touch with our group. As you know, he is the Director of the Mycoplasma Research Institute in Florida.

This hypothesis of a federal mycoplasma virus is consistent with the seven papers (six on mycoplasma) that you published inside the special virus program during 1971 alone.

I think you have an obligation to help us, by immediately agreeing to review the flowchart, and issue a statement, prior to the Royal Society Conference on the Origin of AIDS next month. The Los Alamos AIDS Database now has the flowchart, as well as the first 60 pages of progress report #8 (1971). None of the databases of this federal program are referenced through Los Alamos' supercomputers. The NIH has the flowchart. You can call the archives of the NCI, Judy Grossberg, the librarian, is willing to copy the flowchart and send it to you. She will also confirm the archives can not account for the missing 12 progress reports.

It does appear that many young scientists worked on their specific projects inside this program, truly unaware of the overall diabolical nature of the program, as is graphically outlined by the flowchart. This point is also supported by the 60 page narrative text, leading to page 61, the flowchart. Progress Report #8 (1971).

We believe the flowchart is the "indisputibile" missing link of concrete, direct evidence of a federal virus development program that matches with uncanny precision, the "peculiar epidemiology" of the pandemic of AIDS.

We further believe Batch #751 of the experimental Hepatitis B vaccine that was given in Manhattan will show a "complement relation" with the federal virus. This is also speculated strongly to be true for the smallpox vaccines that were sent to Africa during the late 70's. As smallpox ended because of mass vaccination, AIDS began in mass.

This flowchart and progress reports are the direct, concrete evidence of the "laboratory wellspring" of the genesis of AIDS. According to Colonel Jack Kingston, Chairman of the National Security Advisory Board, the 'Special Virus program is the biggest secret in U.S. history', and 'its' impact on humanity will be greater than the Manhanttan Project'. The AIDS virus appears to be supported by "century long" research.

We truly do need you now. Where is the weak spot in *mycoplasma* penetrans (for example)? We appreciate the deliberate speed that you have shown in responding to our ongoing dialogue. Thank you in advance for your thoughtful reflections on the mycoplasma computer. Where would you like to receive a copy of the flowchart and the first 60 pages of progress report #8 (1971)?

Eagerly looking forward to your replies.

Sincerely,

Boyd Ed Graves, J.D.

Director-AIDS CONCERNS the Common Cause Medical Research Foundation

VIA EMAIL

November 6, 2000

To: Dr. Bob Lee <rboblee@home.net>

From: Boyd E. Graves, J.D. <boyded2001@yahoo.com>

Bob, Thanx for calling. Hope all is well. Can you resend your email on CCR-5? Somewhere I believe I read that there is a racial significance to this HIV entryway. You have sent a lot of (private) email to Alan. I wish you could go back through and send any to me that I might find useful.

McClure in 1972, "Pneumocystic carinii pneumonia in chimpanzees"

In: Proc 23rd Annu Sess Am Assoc for Lab Anim Sci Abstr, October, 1972.

There is also something fishy about the "International Primatology Congress(es)".

The progress reports prove the United States and the Soviet Union were working "side-by-side" in the development of the "Special Virus". The "Special Virus" was not being prepared for an assault on the Soviet Union, they all worked together to seek to "stabilize Africa's population. Remember, our justification for working on these bioengineered systems was for the purposes to be able to 'RESPOND IN KIND'.

It is ludicrous and preposterous that all this bioengineering effort has been directly SOLELY at the Black population, however it is true. The United States nor the Soviet Union would never face a bio attack from Africa. All of this is Bullshit. They were simply carry out the edicts of a White racist, eugenics mindset that Blacks must ultimately be eliminated. Remember, it is Craig Venter, Jr who patented CCR-5 and who has made a mycoplasma with just 300 genes, capable of inducing human cancer.

Black people need more help and more exposure (review) of the flowchart and progress reports of the Special Virus.The flowchart is broken down into five phases and each phase lists the contract and experiments for that section. To me, we should be concentrating on PHASE IV-A.

Somewhere along the line we have all forgotten that HIV was originally called, "Leukemia/lymphoma" virus. Thus, the University of Texas' experiment (72-3262) appears to have significance in that review of their vaccine research from 30 years ago might provide insights for the research being conducted currently. Some of you talk with Garth Nicolson, he promised to get involved (Aug 99). He has not done anything. For the record, CBS showed 'mycoplasma' on its 9/21/00 CBS evening news. They called it the "Armour" of viruses. Is there a chance that someone is in a position to order the videotape or transcript of that broadcast from their archives.

As I mentioned, Dr. Alan Rabson, NCI is attempting to get a copy of the Zinder Report. In 1952, Zinder reported on transduction, or transfer of genetic information by viruses. (Carrying DNA from bacterium to another). I thought earlier this year that our side needed to have a conference. However I have been able to observe how my suggestions are being received. None of you have commented on the NAS Proceedings that conclude the infectious agent for Kaposi Sarcoma has been "Re-arranged" (by man). None of you have commented on the "natural history" of visna. (There is none.)

Dr. Merigan's 1971 paper, "Viral Infections in man Associated with Acquired Immunological Deficiency States", earned him a right to become a consultant for the "Special Virus" SEE page 14, PR#10 (1973). The references to this paper prove ethnic tinkering as far back as 1904. When can we discuss this paper?

Fed Proc 30 1858-1864 (1971).

Without the flowchart, they would win.

Please link in my website: http://www.boydgraves.com

Thanks for all you are doing.

Sincerely,

Boyd Ed Graves

1-888-842-6419

Please note for the record that it appears that Dr. David Hillis, University of Texas is still playing an instrument in their 'orchestrated' silence. The common anscestor (?) of AIDS is VISNA. Visna appears to be man made. It only apppeared in the sheep sent to Iceland. There was no Visna in Halle, Germany or Uzbekistan, the natural stocks for the Karakul sheep. The early AIDS tests in the 1930's on sheep proved that the government needed a six year incubation period for the mycoplasma to have full death effect. It is worthy to note that after complementing the vaccines (small pox to Africa and Hep B to Manhattan in 1978, six years later (1984), they told us what it is.

VIA EMAIL

Date: Friday, November 10, 2000 1:11 PM
Subject: Re: ORIGIN OF AIDS: HISTORY ON HOLD

Dear Eddie and All Whom This May Come:

Greetings,

As is often said, "It ain't over until it is over." An enquiry into the origin of something as evil as AIDS virus cannot be swept under the carpet by a decision of one individual, or one country. Five thousands people are dying every day in Africa due to this man-connected evil disease. If for example, 10 fully loaded Jumbo Jets were crushing everyday in the USA and there was no visible end to the calamity, I wonder who among the judges would dare call that frivolous! And because the calamity is happening in Africa, does that make it frivolous? If it is, then the whole problem of slavery and reparations is frivolous too. Is it?

If a US District court has determined this case to be frivolous, that should be regarded as a mere opinion of an individual who, depending on the degree of his knowledge or ignorance of the facts, passed a judgement that is purely individualistic and barren of all permanent legal truth. There is one more place here in the USA which goes under its generic name of "Supreme Court."

That should be the next port of call. We are assuming that the "nine wise ones" who sit in judgement will be guided by the Divine Order, the true Maker of Africa and her people, the Divine Maker referred to in different African languages as, Mungu, Katonda, Ngozi, Ngai, Nyasaye, Were, etc., to act wisely. Naturally, if the "nine wise ones" chose anything other than wisdom as they have sometimes done when they threw God and His Ten Commandments out of their public schools, there is one more place to go—The International Court of Justice. In short, this matter is far from over and is absolutely not as frivolous as a single opinion of some judge sees it.

We the Knights of Africa in obeyance to the will of "Mungu" will do our best to support any further legal investigations now and in the near future. We urge everyone with a clear mind and spirit to do the same. We reject the theory that this calamity is self inflicted or is nothing but monkey business and monkey originated. We see an apparent connection between early US Special Virus research as presented by Eddy Boyd Graves in his exhaustive research and this out of hand calamity. Finally, we plan to assist with whatever financial resources we can muster.

Sincerely,

Prof. Sir Vincent Mbirika
Sovereign Grand Master
Knights of Africa

December 5, 2000

LeRoy Whitfield
POZ Magazine
349 12th St.
New York, NY 10014

Re: "Precious Thomas Has HIV: Did The Government Give it to Her?"

Dear LeRoy:

Yesterday I had an opportunity to peruse your story. You led me to believe that you were serious in seeking legitimate answers to a high percentage of Black people who believe their government is trying to kill them. We believe the flowchart which you failed to mention represents the "missing link" in the absolute proof of the laboratory origin of AIDS.

You led me to believe that your were going to call several of the experts who have reviewed the flowchart.

You did not.

You had an opportunity to tell the truth and you side-stepped that duty to ensure your individual future success, apparently to the detriment of your own people.

Even further, you cast me as "crazy" in which to demean the evidence, and you did not even speak with our international foundation. In essence, in light of the conclusive information you received during your visit to Cleveland, you found it necessary to continue to ask why, irrespective of the substantial and credible evidence you received

I sincerely believe your article is purposefully false and written to mislead the American people about the concrete evidence of a federal virus development program. The flowchart coordinates over 20,000 scientific papers and a 15 year history of progress reports.

I now believe legal action is the only way to compel you and your magazine to tell the truth. Your gross failure to seek to verify my hypothesis has led to additional death and social rape.

Your actions appear to assist the continuation of AIDS, in a concept consistent with "house negroes" of slavery vintage.

According to her government, "Precious" ain't.

Boyd E. Graves, J.D.
Director-AIDS CONCERNS
the Common Cause Medical Research Foundation
mailto:boyded2001@yahoo.com

VIA EMAIL

December 27, 2000

Dr. Victoria Harden/ NIH Historian

Dear Dr. Harden:

I am responding to an email that states 'no viruses were developed' in the Special Virus program. The program's flowchart ("research logic") proves the United States was seeking to create an immmune suppressing virus. Dr. Alan Cantwell can prove the United States, according to the logic of the flowchart, bioengineered and produced "immune-suppressing" viruses in large quantities. Then we had AIDS in Africa and Manhattan.

It is our position this secret federal virus development program needs to be independently reviewed for the secrets it holds in our current fight with a "special" virus, AIDS. The AIDS virus is "special" because it has an affinity to a blood deficiency marker in the Black population.

The Special Virus program of the United States created and proliferated AIDS as a way to bring "eugenic order" to Africa, consistent with PL91-213 3/16/70. Any independent review of the flowchart and 15 progress reports of this secret Manhattan project will definitively conclude the AIDS pandemic is a direct outflow.

Ultimately, the flowchart document find will be revered as one of the greatest document finds in the history of the world. In light of this secret virus development program, why was AIDS a mystery to your agency?

Your agency did not respond to my January abstract, "A SPECIAL ABSTRACT ON THE MEDICAL ETIOLOGY OF AIDS". The endnotes of this paper contradict the current history conclusions of your agency. Please go to my website: http://boydgraves.com I would be happy to forward you those 15 endnotes and phone numbers of many of the experts who have reviewed the flowchart or written books on the secret program.

Dr. Robert Gallo does isolate a retrovirus in the program nearly ten years before it is announced! This discovery is kept secret to allow the program to continue to move toward a "contagious cancer" that "selectively kills". See pages 104-106, Progress Report #8, (1971). Dr. Gallo "excludes his role" as a "Project Officer'for the AIDS project in his autobiography. Additionally in 1971, "n-demethyl-rifampicin" is identified as an "inhibitor" of the Special Virus.

We will hand deliver petition signatures to the Surgeon General for a review of this secret virus development program. We will ask that Congresswoman Tubbs Jones refer the 3000 signature petitions we left with her to Dr. Satcher. The Special Virus program spent $550 million dollars between 1964 and 1978. Where are the Congressional Appropriations for this "Black budget" program?

The relationship between AIDS and VISNA was established through the program, and it is clear the AIDS virus does ndeed have a mycoplasma trigger for neoplastic transformation. The 'wasting' in humans is identical to the 'wasting' in sheep. We believe we are entitled to have a conference on the true origin of AIDS in conjunction with the flowchart and progress reports of the Special Virus program of the United States of America. Linda Jackson via Dr. Cargill is supposed to be trying to find $250,000 for that purpose.

We have found the wellspring of the genesis of the AIDS pandemic. It is our nation state, it is us.

Please provide us the mechanism to allow our research to be included in the true history of ethnic genetic engineering of the United States of America. According to former President Richard Nixon, the 'order' to depopulate Africa extends only until the end of the 20th Century. See, Weekly Compilation of Presidential Documents, Vol. VI, pg. 734, 3/16/70.

It does appear that 'finally' John Bailar and Iwan Morus are willing to step to the public poodium and debate a secret virus program akin to 'Manhattan'. Our collective effort will deactivate this weapon of racial genocide and cleanse the good hearts of humanity. The Special Virus program met every year at Hershey Medical Center, hosted by Dr. Fred Rapp.

Sincerely,

Boyd E. Graves, J.D.
Director-AIDS CONCERNS
the Common Cause Medical Research Foundation
1-800-257-9837
boyded2001@yahoo.com

February 6, 2001

Dear Garth:

This is from Ed Graves. You have avoided all follow up from our 1999 conference. You led me to believe that you were in the Special Virus program but that you had nothing to do with it.

Dr. Nicolson: AT THE MEETING THAT YOU AND I ATTENDED, I DID INDEED VOLUNTEER TO YOU THAT I WAS A RE-VIEWER FOR THE THE NIH SPECIAL VIRUS CANCER PROGRAM IN THE AREA OF CELL SURFACES AND MEMBRANES. THIS HAS BEEN ON MY PUBLICALLY DISTRIBUTED CV SINCE 1974. I ALSO HELD ONE CONTRACT FROM THE PROGRAM IN THE MID-1970s FOR OUR RESEARCH ON RED BLOOD CELL MEMBRANE GLYCOPROTEINS WHILE I WAS A FACULTY MEMBER (ASSIST. PROF.) AT THE SALK INSTITUTE. THIS HAD NOTHING TO DO WITH VIRUSES OR MYCOPLASMAS.

Dr. Graves: Today I will send you a copy of the tape (AIDS: MIRACLE IN CANADA VHS 1999) of your outbursts at that conference. You have yet to make a public statement on the Flow Chart and this mostly-secret federal virus program that spent $550 million dollars to make AIDS.

Dr. Nicolson: PLEASE SEND THE TAPE. I WAS ONLY INVOLVED IN THE QUITE PUBLIC PART OF THE SVCP. DURING MY TIME ON ONE OF THE REVIEW COMMITTEES ALL INFORMATION WAS PUBLIC. I AM NOT RESPONSIBLE FOR NOR DID I HAVE ANY KNOWLEDGE OF ANY PART OF THE SVCP THAT WAS KEPT FROM PEER-REVIEW ADVISORY COMMITTEES.

Dr. Graves: It is Dr. Len Horowitz who casts you as 'Gallo's counterpart' in his book. I will be happy to provide the citation, page 506 "Emerging Viruses: AIDS & Ebola"

Dr. Nicolson: PLEASE PROVIDE A COPY OF THE REVELENT PAGES. I STRONGLY DOUBT THAT LEN CAST ME IN THAT LIGHT, BECAUSE I HAVE ALWAYS BEEN QUITE CRITICAL OF THE VARIOUS RETROVIRUS LABORATORIES AT NIH AT THE TIME.

Dr. Graves: Your 1972 citation classic is cited by Luc Montagnier as the 'second greatest scientific paper of the 20th Century'.

Dr. Nicolson: THIS PEER-REVIEWED PAPER IN SCIENCE WAS ON THE FLUID MOSAIC MODEL OF MEMBRANE STRUCTURE.

Dr. Graves: See- Singer, S.J. and Nicolson, G.L. The fluid mosaic model of the structure of cell membranes. Science 175: 720-731 (1972). You have been working with the structure of mycoplasmas for 30 years.

Dr. Nicolson: NOT TRUE. I HAVE ONLY BEEN CONDUCTING RESEARCH ON MYCOPLASMAS SINCE MY WIFE ALMOST DIED OF A MYCOPLASMAL INFECTION IN 1987. MY FIRST PUBLICATION ON MYCOPLASMAS WAS IN 1995. PLEASE FURNISH YOUR EVIDENCE FOR SUCH A RASH STATEMENT.

Dr. Graves: Nicolson, G.L. and Nicolson, N.L. Doxycycline treatment and Desert Storm
JAMA 273: 618-619 (1995).

Our collective re-work of your experiments will shed multitudes of information for our current advances against this federal virus. The Flow Chart reveals with specificity the significance of your work inside the program. Your colleagues

(Dr. Basil Wainwright and others) have reviewed your work and concluded as is properly represented in the 'history' chapter of my book.

Dr. Nicolson: WHAT COLLECTIVE REWORK? WHAT EXPERIMENTS? WHAT COLLEAGUES? DR. WAINWRIGHT IS NOT MY COLLEAGUE, AND TO MY RECOLLECTION I HAVE NEVER SPOKEN TO THE INDIVIDUAL ABOUT THIS TOPIC.

Dr. Graves: I accept that you think there may be a factual error and on each point I will make available the citation. Between 1972 and 1978 you are affiliated with the development of AIDS.

Dr Nicolson: I WOULD LIKE YOU TO PUBLICALLY STATE YOUR EVIDENCE FOR THIS, AND I CHALLANGE YOU TO PROVIDE SUCH EVIDENCE.

Dr. Graves: Suddenly you left Salk and crept under the radar at Irvine. However, you state your tenure with Irvine differently than they. One of you is wrong.

Dr. Nicolson: SUDDENLY I LEFT THE SALK INSTITUTE? I LEFT THE SALK FOR A PROMOTION TO FULL PROFESSOR AT U.C. IRVINE. MY GRANTS (ALL IN THE AREA OF CANCER) WERE TRANSFERRED TO U.C. WHILE AT U.C. IRVINE I CERTAINLY WAS NOT "UNDER THE RADAR." I CONTINUED TO MAKE PROGRESS ON MY ACADEMIC CAREER, AND THIS IS ALL IN MY CV.

Dr. Graves: You left Irvine on 6/30/80 heading for (MAX PLANCK west) "KING COTTON", M.D.ANDERSON. While the Special virus was "complemented" to African and Manhattan vaccines, you assisted the process of the development of the incapacitating mycoplasmas by your silence on the secret work on Texas prisoners emanating from the "overabundance" of former German scientists.

Dr. Nicolson: MY WORK AT THE UNIV. OF TEXAS M. D. ANDERSON CANCER CENTER WAS ENTIRELY IN THE AREA OF CANCER METASTASIS UNTIL THE NEAR FATAL ILLNESS OF MY WIFE IN 1987. AT THE TIME SHE WAS A FACULTY MEMBER AT BAYLOR MYCOPLASMA PROGRAM. BY FINDING OUT WHAT WAS WRONG WITH HER, WE HAVE BEEN ABLE TO HELP THOUSANDS OF PATIENTS RECOVER FROM POTENTIALLY LIFE THREATENING DISEASES. BUT WE STARTED ACTUAL RESEARCH ON MYCOPLASMAS AFTER MY STEP-DAUGHTER RETURNED FROM THE GULF WAR 101st ABN DIV AND CAME DOWN WITH GULF WAR ILLNESS AND WASHED OUT OF ARMY PILOT TRAINING AT FORT CAMPBELL. OUR WORK WITH TEXAS DEPT. OF CRIMINAL JUSTICE EMPLOYEES STARTED WHILE WE WERE HELPING EMPLOYEES WHO ESSENTUALLY HAD GULF WAR ILLNESS. WE FOUND THE SAME MYCOPLASMAL INFECTIONS IN THESE EMPLOYEES AS WE FOUND IN GULF WAR ILLNESS PATIENTS. SPOUSES AND MOTHERS OF THESE TDCJ EMPLOYEES LATER DUG UP THE DOCUMENTS PROVING THAT MYCOPLASMA TESTING WAS TAKING PLACE IN THE PRISON SYSTEM AND BAYLOR COLLEGE OF MEDICINE WAS INVOLVED. THIS WAS THE SUBJECT OF A UPN NETWORK NEWS BROADCAST (OUT OF SAN ANTONIO) THAT WE WERE IN LAST FALL.

Dr. Graves: Although you have peer reviewed papers on your early mycoplasma work, your discoveries have largely gone unnoticed.

Dr. Nicolson: WHAT PEER-REVIEWED PAPERS ON EARLY MYCOPLASMA WORK??? THIS IS A COMPLETE FABRICATION BY YOU.

Dr. Graves: It is your peers whom I quote, who say your work inside the Special virus program is relevant to the destruction of the Black Population by the year 2066.

Dr. Nicolson: MY WORK IN THE SVCP WAS EXACTLY AS I DESCRIBED IT. I WAS A REVIEWER ON SVCP PROPOSALS IN MY AREA OF EXPERTISE, MEMBRANE STRUCTURE AND ORGANIZATION. IF ANYTHING, OUR RESEARCH HAS DONE MUCH TO

SAVE THE BLACK POPULATION FROM COINFECTIONS RELATED TO AIDS. IF YOU WOULD BOTHER TO CONTACT MY COLLEAGUE, DR. RICHARD NGWENYA OF THE JAMES MOBB AIDS CLINICS IN ZIMBABWE (jamesmob@africaonline.co.zw), CLINCS THAT WE SUPPORT FINANCIALLY, I AM SURE THAT YOU WOULD FIND OUT QUITE THE OPPOSITE OF YOUR STATEMENTS.

Dr. Graves: We say your 1972 project at Salk Institute shows scientific research beyond that which is recorded for that period in time.

Dr. Nicolson: WHAT RESEARCH BEYOND WHAT WAS RECORDED? THIS IS A COMPELTE FABRICATION. WHERE IS THE EVIDENCE?

Dr. Graves: We can show that your work in the Special virus program is related to the stabilization of a mycoplasma for characterization for large scale production. (Progress Report#9 (1972))

Dr. Nicolson: THIS IS RIDICULOUS. I NEVER DID ANY WORK ON MYCOPLASMAS UNTIL THE LATE 1980s, I HAVE NEVER WORKED ON LARGE-SCALE CULTURING OF MYCOPLASMAS. THE ONLY CULTURING OF MYCOPLASMAS THAT WE HAVE DONE HAS BEEN TO GROW SMALL QUANTITIES TO SERVE AS CONTROLS FOR OUR MOLECULAR TESTS.

Dr. Graves: You wanted to help the process in 1999, however we have not heard from you until now.

Dr. Nicolson:ONE CAN CERTAINLY SEE WHY. WHO WOULD WANT TO BE ASSOCIATED WITH HALF-TRUTHS, FABRICATIONS AND MISSTATEMENTS.

Dr. Graves: Your early work is critical for the medical advances necessary for the deactivation of this federal virus.

Dr. Nicolson: OUR WORK IS MAINLY ON THE CO-INFECTIONS THAT ACTUALLY CAUSE THE MORBIDITY IN AIDS PATIENTS. MY WIFE'S RESEARCH IN THE AREA IS PUBLISHED AND SHOWS HOW THE AIDS VIRUS INCORPORATES INTO THE GENOME TO CAUSE CELL DEATH. THIS IMPORTANT WORK WAS RECENTLY PUBLISHED.

481. Nicolson, N.L. and Nicolson, G.L. Nucleoprotein Gene Tracking: localization of specific HIV-1 genes to subchromatin nucleprotein complexes in HIV-1 infected human cells. J. Cell. Biochem. Suppl. 32: 158-165 (1999).

485. Nicolson, N.L. and Nicolson, G.L. HIV-1 genes are localized in specific nucleoproteins in subchromatin complexes in HIV-1 infected human cells. Int. J. Med. Biol. Environ. 28(1): 25-31 (2000).

Dr. Graves: Please recall that Richard Nixon outlined that the Special virus was projected only until the end of the 20th Century. See, PL91 -213, 3/16/70.

With you, working with Dr.Goosby/Rabson/Cargill/Snead/Cantwell/Horowitz we can review the Special virus program, and if people with HIV/AIDS are assisted, we will have sought to better serve the natural order of mankind.

Please join our effort with your vigorous support. I am at 785-263-1871. I will forward to you the questions and answers in our ongoing dialogue with Dr. Rabson and Dr. Goosby.

The 30 year old Flow Chart proves the United States actively sought to thwart the Black population by creating, producing and proliferating a "special" virus that depletes the immune system. Via overnight mail I am sending you a full-size copy of the Flow Chart. If you would like I can also send you a videotape of the international conference.

Let the review process begin with an evaluation of your project and papers for the 'special' virus. You have so many wonderful people who love and appreciate all you are doing. Will you stand up against the mycoplasma that is targeted at Black people?

Dr. Nicolson: UNFORTUANTELY, THE MYCOPLASMAS THAT WE HAVE BEEN WORKING DAY AND NIGHT TO IDENTIFY AND FIND THERAPEUTIC SOLUTIONS FOR PLAY AN IMPORTANT ROLE IN GULF WAR ILLNESS, CHRONIC FATIGUE SYNDROME, FIBROMYALGIA SYNDROME, RHEUMATOID ARTHRITIS, ALS, MS. AIDS, LUPUS AND MANY OTHER CONDITIONS. WE ARE DOING EVERYTHING THAT WE CAN TO HELP PEOPLE RECOVER FROM THESE CHRONIC AND POTENTIALLY FATAL DISEASES. THIS EFFORT IS NOT SOMETHING THAT I TAKE LIGHTLY.

Dr. Graves: Garth, I am looking forward to speaking with you again, and truly working with you to jointly attack this federal AIDS program. Thanks for all you contiiinue to do for humanity.

Sincerely,

Boyd E. Graves, J.D.

cc: Don Scott

(above) "INAUGURATION DAY" With the 1971 U.S. Special Virus Flow Chart in the foreground, the Bush inaugural in the background, and Boyd E. Graves, J.D. in the middle - January 20, 2001 Boyd E. Graves J.D. shares the Special Virus Flow Chart with Texas radio audiences on the one year anniversary of the U.S. Department of Justice naming the office of the President of the United States as lead defendant to his petition for Global AIDS Apology. George W. Bush inauguration plays live on the television inside the engineer's booth. Dr. Graves was originally scheduled for a two hour live interview, but the show was preempted by the network's coverage of George W. Bush's inauguration ceremonies. The Sixth Circuit Court of Appeals in Cincinnati dismissed Dr. Graves' AIDS case as "frivolous", as predicted by Dr Graves, on Election Day November 7, 2000. The world will never forget the Election Day chaos which ensued.

VIA EMAIL

February 7, 2001

Dear Garth:

You pontificate as if somehow I can't read or I can't find your name or work in a manual that identifies with absolute accuracy your early work to the detriment of the Black population.

1. Any dim wit can read your 1972 paper (citation classic) and discern that you are one of the top mycoplasma experts in the world. It is ludicrous for you to assert you only began work on mycoplasmas in the 1980's when you make comparisons to the structure of mycoplasma in your citation classic. It is you Dr. Nicolson who informed the Enquirer Magazine that HIV and Gulf War Syndrome are a lethal and incapaciting mycoplasma.

2. Your tenure as a full Professor at UC Irvine was without pay. There is a discrepancy between your CV and the listed duration of your government assignment at UC Irvine. Wasn't Eric Traub working there at some point? Len Horowitz believes so.

3. Look your name up in Len Horowitz's index "Emerging Viruses". I am a Black man who can read. Case in point:

I direct your attention to page 294 of progress report #9 (1972).

In 1972 you undertook the following project to the detriment of the Black population.

Project 3. (Dr. Nicolson) "The Amount and Distribution of Cell Surface Antigens in Normal and Tumor Cells Will Be Studied Using Radioimmune Assay, Cytotoxic Methods and Epecially EM Visualization Using Ferritin Conjugated Antibodies and Plant Lectins."

In 1973 you provided a summary of your project and seven scientific papers effectively allowing for the immune system of the Black Population to come under attack. Your name is affiliated with the program from 1972 - 1978 until you left Salk for a non paying full professorship!

You sound as if you can not read. Your pompous posture implores you to boast of seeking to take some legal action, but we all know your work is instrumental in the devastation of the Black population.

You arrogantly boast of having some affiliation in some African clinic, list the African American Doctors who have peer reviewed your 1972 citation classic.

We believe it makes sense for the Surgeon General, Dr. Goosby and Dr. Rabson to know more about this federal virus program. Do you support our basic call for an immediate independent review of this federal virus program that spent $550 million dollars? See, Table One, Funding History, Progress Report #15 (1978).

We are prepared to meet each of your meager reasons for non-review of this federal virus program outlined in 1970 in PL 91-213. Place on the record your review of the Flow Chart. Place on the record your review of the conclusions of the National Academy of Sciences, AIDS "evolved from the man-made Visna virus". See, Proceedings, NAS, Vol 92, pp. 3283-3287, (April 11, 1995).

The prevailing understanding in opposition to the premise this program was a search for a cure for cancer lies in the fact that you and others in tumor metatasis have long been aware and a follower of Dr. Otto Warburg. Dr. Warburg's early paper on the 'cause and prevention of cancer' was equally suppressed as have been the Flow Chart and your paper. The Special Virus can not argue they were searching for a cure for cancer in light of Warburg's work and Thomar's.

On page 24, PR#8 (1971) a cure ('"inhibitor of replication") is announced. Do you support our call for a review of the drug n-demethyl rifampicin as to the immediate medical breakthroughs this drug may provide? Finally, it is the 1971 human RDDP located by Dr. Gallo (id at 104- 106) that is co-mingled with mouse, sheep, monkey, cow and other animal viruses to best explain the sudden emergence of the two primary strains of HIV.

We will continue to provide our references and expect proper public recognition as we truthfully re-examine this one hundred plus year development of an ethnic genocide biological program to the detriment of the U.S. Constitution and the will, heart and soul of the American people. I await your apology.

Sincerely,

Boyd E. Graves, J.D.

Director-AIDS CONCERNS

the Common Cause Medical Research Foundation

boyded2001@yahoo.com

PURE BLOOD, INC.

pureblood1@juno.com 607/363-2264
P.O. Box 521, Downsville, N.Y. 13755

Sect. General Kofi Annan
The United Nations Headquarters
New York, N.Y. 10017

May 10,2001

Dear Sect. General Annan:

According to The New York Times, Africa has about 70% of the worlds 36 million people infected with H.I.V., the virus that causes AIDS and 17 million Africans have already died of the disease. However the Times also reports "AIDS drugs now extend survival time fourfold". Therefore, pharmaceutical companies can now depend on four times their normal income from these drugs, consequently your efforts to secure additional funding for AIDS victims is time well spent; that is if you accept the objectives of the CFR, i.e. making lots of money on the Nazi Visna virus and reducing the population growth of pre-selected human beings.

Surely, your office is aware of the origin of AIDS. Mr. Boyd Ed Graves has already brought charges against the U.S. Government, several M.D.s have written books on the subject including Dr. Leonard Horowitz – EMERGING VIRUSES, AIDS & Ebola, NATURE, ACCIDENT or INTENTIONAL.

My colleague Dr. Basil Wainwright has already cured over 1,000 AIDS victims to PCR Undetectable in Kenya (Encl. 1), but the CFR and the pharmaceuticals have managed to keep his technology quiet. While Dr. Robert Gallo has over 1,900 Patents on the creation of the HIV virus, Dr. Wainwright has the only U.S. Patent for the cure of AIDS (Encl. 2). Leaders of several Nations in Africa are anxious to establish Dr.Wainwright's Polyatomic Apheresis Clinics. May we request that you (not the pharmaceuticals) consider funding his clinics that have been so successful in bringing AIDS, Malaria, Sleeping Sickness, Hepatitis C, etc., etc. to remission without any negative side effects.

Sincerely Yours
Pure Blood, Inc.

Gerald O. Rennerts
President

c.c. Dr. Vincent Mabirika
Knights of Africa

Mr. Boyd Ed. Graves

Dr. Basil E. Wainwright

PHASE III - A

TRANSCRIPTS

www.boydgraves.com

Dr. Graves' speech "The Societal Impact of the Special Virus Program of the United States of America" was presented at the First International Conference on Degenerative Disease hosted by the Common Cause Medical Research Foundation, in Ganonoque, Canada, on August 21, 1999. Many members of the international medical and scientific communities were in attendance, including esteemed Gulf War doctor and Special Virus program scientist, Dr. Garth Nicolson. Dr. Nicolson interrupted this speech when Dr. Graves displayed the 1971 Special Virus Flow Chart and the accompanying Progress Report by stating, "My name might be in one of those (Progress Reports), but I had nothing to do with this (the creation of AIDS)." The world's first glimpse of the U.S. government's secret AIDS Flow Chart was at the same moment in time immediately authenticated by one of the participating scientists who happened to be in the audience this fateful day.

{See PHASE V: "AIDS: MIRACLE IN CANADA" VHS 125 minutes}

The Societal Impact of the Special Virus Program of the United States of America

by Boyd E. Graves, J.D.

To the people of the world:

It is indeed a unique honor and a rare privilege, to have this wonderful opportunity to recapitulate, even perhaps, regurgitate, my thoughts, reflections, foresight and conclusions here today, in my capacity as Director-AIDS CONCERNS for the international medical research foundation, Common Cause.

However, before I begin, it is absolutely imperative that I make mention of a host of caring, concerned, dedicated, freedom-loving people, from around the world, who have unselfishly given of themselves, in so many ways, for the reality of my presence here today.

First let me say thank you to those pioneers in this struggle, this global war, who are no longer with us. Thank you mom and dad. Thank you Ted Strecker, Thank you Jacob Segal and Thank you Douglas Huff. I would also like to say a special thanks to Zears L. Miles, Jr., in who's reincarnation I blossom. Were Zears still living, it would be he standing here today.

All of these individuals, in one way or another, have given unselfishly to the cause of awakening a sleeping tsunami of public outrage, over a virtually unspoken, stealth world policy of population control. A policy designed to cull humanity. A policy designed to kill humans in order for future humans to procreate freely. It is absurd. It is insane.

It is the ultimate crime, 'state-sanctioned premeditated murder'.

What is the true nature of our society that so heavily punishes indiscretions and misdemeanors, while fostering hidden programs that kill and debilitate millions of innocent people. We maintain disgust with the population control agendas of the 1940's, we must now collectively direct our outrage to the doorstep of the perpetrators. History will ultimately show the heinous nature of the warped mentality that governed the last half of the twentieth century.

Along this personal journey in search of thrush and fact, there have been a number of individuals who have continually propped me up with their expertise, research, unfettered love and support. Some of them are very familiar to you; Drs. Strecker, Cantwell and Horowitz. However, many other are not known by you at all. Thank you Janet, Maureen, Mary Jane, Candace, Angie and Sheila. Thank you Vincent, Eric, George, Mike, Terry, Ken, Brother Jabbar, Kwame and Tom, and a host of others too numerous to mention.

But for the efforts of these magnanimous individuals, I would not be here today.

Second, I would like to make a few remarks about my coming to know Don Scott and the Common Cause Medical Research Foundation. One day a lady called and asked had I heard of a book entitled "Brucellosis Triangle". I said I had not and she said 'well you can't find it in the United States, you have to call to the author and get it through his small publishing company'. Shortly thereafter, I was on the telephone with a soft-spoken man whom seemed to radiate peace and tranquillity through the phone. Don and I have spoken a number of times since that first phone call and I always feel as though he has the voice of an "off-season" Santa Claus.

However his soft-spoken speaking voice masks his true nature, a relentless warrior with dignity, in search of truth and fact. As seemingly varied as our collective pasts might initially appear, we share a common cause. We also share a common purpose as is evidenced by our assembly here today.

We gather here in search of an answer to a nagging scenario that continues to plague the great scientific and intellectual minds of humankind, across this planet. Is there an interrelated natural or synthetic origin of a number of neurodegenerative/systemic degenerative diseases. My research concludes the answer is yes.

I seek a resolution from this Conference calling for an independent global review of the schematic diagram, the 'research logic flow', of the "Special Virus" program of the United States of America. The schematic is listed as page 61 of Progress Report #8 (1971) of the "Special Virus" program of the United States. To the people of the world, I present the indisputable scientific logic that lead to the creation, production and proliferation of AIDS.

Let me say forthwith, that the governments of the United States and France have conceded that their program in search of a cancer recipe and an eventual cure for cancer, did not begin until 1984. This schematic diagram is not a diagram in search of a cure for cancer. This is a schematic diagram that seeks to create a "special virus" which affects the immunological system of humankind. I direct your attention to the July 29, 1999 Nature journal and the scientific paper submitted jointly by the Whitehead Institute for Biomedical Research in Cambridge, Massachusetts and the Institute Pasteur in Paris, France.

Without an independent global review, humanity will be forever subjected to the intentional false propaganda that purposeful and directly supports the crime of state-sanctioned premeditated murder. I ask for the full support of this Foundation in seeking a definitive, independent global review of the Research Logic Flow of the "Special Virus" program of the United States of America. I am of the opinion of the "Special Virus" can be put together, it can be taken apart. This premise is further supported by the schematic itself. Under Phase IV-A, IMMUNOLOGICAL CONTROL, the "Special Virus" was shown to be controllable.

Prior to mass production, control of disease/virus replication had been shown in animal models and (human) tissue cultures. Accordingly, we have within the 15 progress reports, the contracts for this and every other segment of a massive man hunt for a virus that would selectively kill. Before I go further it is important to note that the archives of this "hidden" federal program, located at the National Cancer Institute (NCI) in Bethesda, Maryland maintains only three of the fifteen reports.

It is also important to note that the "Special Virus" program met every year at Penn State University from 1963 through 1978 and the Hershey Medical Center has no records of any such meetings. There can be but one conclusion, the schematic diagram is real, and the progress reports do indeed represent the missing data, experiments and contracts, that were the result of nearly a century long search for a population control weapon. Recently Don sent out a 'biological warfare timeline'. It is significant that 'nothing' is listed between 1957 and 1969.

The record reveals that during this period the United States began its first official program to hunt for an offensive biological weapon. The program was called "Special Operation -X (SOX) and was conducted by the joint efforts of the military, Central Intelligence Agency, pharmaceutical companies and countless numbers of collegiate microbiology departments around the United States. Since the inception of the US's biological warfare program in the 1940's, it is clear the program co-mingled defensive and offensive research into incapacitating and lethal agents. In accordance with Progress Report #8, (1971) at page 276, we find definitive proof that the United States made a conscious decision to begin the "Special Virus" program in 1962. The very first thing the United States undertook to accomplish was the inoculation of newborn "nonhuman" primates. The monkey inoculations began on February 12, 1962. However, any independent review of page 289 might conclude the monkey inoculations began in 1957. These animals were inoculated with every conceivable human virus known to humankind. The report of the inoculations is listed at pages 276 - 289 of Progress Report #8 of the "Special Virus" program of the United States. This Conference must seek to further make public, 'these specific experiments' in that they do provide compelling scientific evidence of the 'man to monkey' etiology. The iatrogenic origin was independently confirmed earlier this year in a scientific paper presented in Atlanta, by the Instate for Virus Research, Kyoto University, Kyoto 606-8507, Japan. It is also true that many of these inoculated monkeys were subsequently released back into the wilds of Africa.

As the "Special Virus" program rolled along during the 60's, it was not until April of 1969 did the program sponsor a Conference under the auspices of Fort Detrick. The Conference was entitled, "Entry and Control of Foreign Nucleic Acid". I also understand a Conference was held in May, 1969 and then we have Don's portrayal of the notification of Congress on June 9, 1969. Although Horowitz believed it occurred on July 1, 1969. It did occur. The United States under sworn testimony to the US Congress admitted it had been working on a synthetic biological agent that would lead to world wide scourge and black death type plague, in certain geographical areas and regions of the world. In my personal assessment of the factual record to date, the United States had isolated the AIDS virus. By the US government's own admission, the National Academy of Sciences, National Research Council had begun their research in 1967. Oddly enough this is the same year the Central Intelligence Agency began development of its 'poison dart gun', the microbioinoculator. Are we not now entitled to a full accountability of the usage and whereabouts of these assassination tools? Are we not now entitled to an accountability of the 60,000 liters this hidden program produced in 1977? These are serious, disturbing questions in need of answers which appear to reflect another US government policy hopeless out of control, to the detriment of humanity.

Boyd "Eddie" Graves joined the U.S. Navy in 1970. While in the Navy he was the only midshipman to serve on the Brigade staff the entire year. At the Naval Academy he studied Mandarin Chinese as part of his curriculum and was placed in charge of the nuclear missile launch codes while serving as the Communication Officer aboard the missile destroyer U.S.S. Buchanan. In November 2000 Navy Historian Robert Schneller interviewed "Eddie" as part of his new book on race and the U.S. Navy, "Breaking the Color Barrier: Racial Integration at the U.S. Naval Academy." The first part of Mr. Schneller's' two part interview transcript is reprinted here, with permission.

U.S. Naval Historical Center Oral Histories

Boyd E. Graves
United States Naval Academy Class of 1975

17 November 2000

Robert J. Schneller, Jr.
Contemporary History Branch

Schneller: This is Bob Schneller of the Naval Historical Center and I'm here with Eddie Graves, Class of '75, in Nimitz Library, to do an oral history for *Breaking the Color Barrier*, and as is the norm, I'll just go ahead and start from the beginning. What's your birthday and where were you born and raised?

Graves: Yes. Let me kind of give a kind of an overview and if it encompasses some of these early questions, that'll kind of all fit in together.

Schneller: Good.

Graves: Boyd Eddie Graves, born July 10, 1952. Born and raised in Youngstown, Ohio, northeastern Ohio. I am African American. My parents are both African American. My mom earned her college degree at the age of 16. My dad came from a steel/coal-miner West Virginia environment and had an educational level in the 10th grade, I think, up to the 10th grade.

I'm the third of nine children. I have two older brothers, so it was difficult having two older brothers because they were of different sizes, height, and my eldest brother is similar to my height, so I immediately got those clothes to grow into and then my middle brother, between me and my older brother, was a little bit larger and taller, so those clothes were waiting for me once I became a little bigger, so early on I kind of knew that, boy, I'm not going to be getting any new clothes and that never became a phobia for me. That, goodness, you've got two older brothers. They're of different heights. Clearly, you're just going to grow into their clothing and wear their clothing, so that never was a barrier for me.

As a child, probably stealing from my mom's educational experiences and background, I have always had a propensity for excellence in academics. As a child, and following that same trend, my family had an association through the church, the Baptist church, and there were others there who had college education and degrees, and I was highlighted vis-a-vis almost like a child prodigy, where I was taken probably about the age of 9, 10, or 11, somewhere in that range, and tested for intelligence quotient, and I think it went far beyond their normal testing procedures, because there were some group sessions with three or four white adults behind a table and me on the other side, asking questions that required more than a single answer. They required a sentence or two or a paragraph or two answer, and something very strange happened that evening.

It probably wasn't until then that I knew that I was a pretty smart guy from our assessment of knowledge as we knew it then, and even coming home from that I.Q. testing at that age, the gentleman who had taken me there, who has since passed, Mr. Hugh Frost, saw a Dairy Queen up ahead as we were coming back home. We lived in northeastern Ohio and we were in western Pennsylvania, I believe, Westminster College at that time, and he says, "Oh, there's a Dairy Queen." He says, "I'm going to stop at the Dairy Queen," he says, "and Eddie, you can have anything you want," so apparently the evening had gone well and I had felt pretty comfortable in answering many of the questions and I knew that I'd darn near gotten virtually all of them right that evening.

So I went on from a high school standpoint to join the Boy Scouts, became an Eagle Scout very early on, entered the Order of the Arrow, which is a kind of removed upper level for the normal Boy Scout program. Won some debating contests, oratory contests, joined the Debate Club, of course.

Schneller: Was your neighborhood integrated or was it mostly black, and were the schools that you went to integrated or mostly black?

Graves: Our neighborhood was integrated. The Hungarian neighbor, Bruce Joseph, about my age, living right next door. Right across the street, the Hungarian-Polack-Catholic Church. Just down the street the black Baptist Church.

Schneller: Was that your church?

Graves: No, we went to another one.

Schneller: What was the name of your church?

Graves: Our church was Jerusalem Baptist there in Youngstown. But in our neighborhood itself, around the corner my German neighbors had a boy my age, Wally Kulick. My high school class, 1970, from East High School there in Youngstown, the last high school class where there were more white than black in the graduating high school class. After 1970, the following classes were all more black than white.

Schneller: Were you in '70 or '71?

Graves: '70. My high school class being the last class with more; higher percentage of whites than blacks, and that high school still; it's no longer there today, but I went on to become the class president of my high school class in 1970, and academically was fourth in the class and Chess Club president. What I'm saying is that from the time period when I was tested very early on, almost like a child prodigy kind of thing, that I continued in that vein academically and scholastically, if you would, and that led me to the Academy, choosing the Academy, primarily because I had four brothers and four sisters. My parents; my mom, at the time that I was growing up, was just a homemaker and after the rest of the brood came out of the batch, she then went back to teaching, as a school teacher, and she taught German in the elementary school system of the junior high school system there in Ohio, and by bringing all that together, I guess it's like how do you come to choosing the Naval Academy back in a time period when that wasn't fashionable.

Schneller: Before we get there, did you have a job or jobs while growing up and were you expected to contribute part of that income to the family? I would imagine it would have been tight with nine kids.

Graves: It was tight with nine children and yes, any little job would be helpful, I think. As a very, very little boy, I wanted to be like the big kids and there was; this is so silly. There was a watermelon stand, in fact, right in the next block, where, when the watermelon truck came in, you would have a line of workers pitching watermelons [laughter]. This is northeast Ohio nonetheless, but when the watermelon truck came in, hey, a penny a watermelon to catch one and pitch it to the next guy. You see what I'm saying? So, I was too little to get in on that, but it seemed like such a fun kind of job. That camaraderie of working in the watermelon line.

Schneller: How many people were in the watermelon line?

Graves: Five or six.

Schneller: I take it was a bucket brigade type situation.

Graves: It was.

Schneller: Passing, tossing watermelons.

Graves: One off the truck to the guy down below to the next guy to the next guy to the next guy as he's stacking them, so this procedure's going on. That's how they, boom!, they stack the watermelons, so a penny a watermelon you were paid per se, and a penny a watermelon you were deducted if you dropped one, nonetheless, kind of thing, so I longed for that position, but I really don't think I; I maybe did it but without pay eventually.

Schneller: How old would you have to be before you could—

Graves: I think it was mostly on size, and certainly the older watermelon handlers wanted to test you by tossing one with a little bit more force, if you would, to see if you're going to— [laughter].

Schneller: Medicine watermelons.

Graves: Sure, why not? But from a job standpoint, we did the normal; we delivered papers, newspapers. Sundays were hectic because it was always a little larger and a little heavier to drag the bags, if you would. The dogs were always a problem as a newspaper kid, newspaper boy. The challenge of getting up early to get out to pick up your newspapers, to deliver them on time and that kind of thing, and even through the cold and the snow. That sense of—

Schneller: Which there was a lot of in Youngstown.

Graves: Absolutely. We're not too far removed from the lake effect. Cleveland gets dumped on really handily and people are really surprised by the amount of snow that Cleveland gets, but we weren't necessarily spared because we were some 60 miles away from Cleveland per se, but that newspaper job was tough and because they were heavy and, as I said, Sundays we were out there pretty early, but it was a job.

Schneller: Did you have that all through high school?

Graves: No, I moved on, I think, maybe like 10th grade to; I'm not sure if you're familiar, you might be, with Isaly's Dairies. They're prominent in Ohio. I don't know if they spilled over to Pennsylvania or not, but Isaly's Dairy, I was like a counter help person, putting in the barrels of ice cream that would come in and open them up and priming them and removing the trays from the deli and taking them back into the freezer or to the refrigerated storage and cleaning up and closing the store and opening the store and that kind of thing, so that kind of carried me through high school, mixed in with my activities of the Boy Scouts. We were pretty prominent back then and pretty competitive. Had a wonderful Boy Scout troop so there was an intermingled mix of activity.

Schneller: What sports and extracurriculars did you participate in in high school?

Graves: Sure. I'd mentioned Debating and Chess Club and believe it or not, I was also; had tried track and I remember getting left behind in the dust in one relay race where we were at a visiting track that had a section that like went under bleachers, so there was a point of time where you were out of view of the other sections of the track, and I received the baton and this was the home track for the other team, and their guy got the baton and knew that section very well and he came out of the chute and everybody's saying "Where's, where's, where's Eddie?" It was almost as if I was standing still. So I knew track wasn't necessarily my forte, but I didn't do football. I was a little short and a little small for that.

Schneller: And that was big football territory.

Graves: Oh, absolutely. Still is. So, you felt less as an athlete because I wasn't a football player, so the academics was really important to me.

Schneller: Was it always assumed that you were going to go on to college and is that just because of your mother's drive, your mother's career in education, or because you were identified as a prodigy?

Graves: I don't know if it was guaranteed that I was going to go on to college.

Schneller: Did your parents stress education and going on to college as being important or did they just want all the kids to finish high school?

Graves: There wasn't a lot of stressing going on, not to me. They may have concentrated on the other children in the family who weren't so academically inclined, and I don't think I ever got lectured about academics because there wasn't any need. I was pretty much a straight A student through grade school and junior high and high school.

Schneller: So you could say it was assumed that you were going on to college, then?

Graves: There was no assumption to go on to college. My two brothers didn't, per se, help me and there was no demand. There was no stress with respect to that.

Schneller: What did your parents teach you about race and what sort of attitudes towards race did you observe in Youngstown while growing up? Did you encounter much prejudice and so forth?

Graves: I don't think I encountered very much prejudice as a child. My best friend was Hungarian Catholic next door, and boy, we hung out all the time. My other best friend around the corner was immigrant German and we hung out all the time and I grew affection for my friends and they were my closest friends, my white friends, and there was never any teaching on race. There was no emphasis "the white man this or any of that" back then. We're talking late '50s, '60s. Not in our family, and I think that had something to do with the fact that we were in an integrated neighborhood.

Schneller: And not one subjected to white flight. It sounds like it was pretty stable throughout your youth.

Graves: Through my time period, but as I was saying, the white flight was beginning to take place as I identified that just earlier that—

Schneller: The high school graduation.

Graves: Right. As was noted by the high school scenario that we were the last class to have a majority of white students, and that's very important to note because in running for class president for 1970, my opponent was an African American who espoused a more radical approach, more racist approach, apparently for some reason or another, in wanting more black influence or African American influence or Negro influence, whatever we were calling ourselves at that point, and indeed, it was the white folk that got me elected.

Schneller: Because you were more moderate and that appealed to them?

Graves: I was just me. I was just me. I didn't buy into the race stuff like that. I thought it was detestable quite frankly, but I didn't highlight it.

Schneller: How did you decide to go to the Naval Academy? What fueled your ambition to become a midshipman?

Graves: You know, I'm not exactly sure what the impetus was, and I thought about this many times back ago, what was it, what were you doing back then that made the Naval Academy stick out? My dad had served in the Navy back in the '40s as an enlisted, so—

Schneller: What did he do?

Graves: I'm not exactly sure, but I know that he had served in the Navy, and so in that sense, I knew Navy over "Army." I had labored through my Boy Scout mile swim. If you talk to some of my compradres here today from the home town, they'll tell you that, "He's the only guy I know that dog paddled his way for the whole mile swim" [laughter]. I was that determined to make it, though, so over time, I certainly learned how to swim. I'd had an incident earlier in life where maybe about the age of seven or eight, I was visiting West Virginia and visiting relatives, and by an odd set of circumstances, someone had pushed me into a river with a pretty good current, the Kanawha River down near Charleston, West Virginia, and I was under going down stream and a white guy down at the next boat dock saw what had happened and just like the, kind of like the Superman scenario, taking off his suit jacket, kicking off his shoes. He dove in from the next boat dock to intercept me and save me and got me to the surface, so I had a phobia of water [laughter] certainly after that, but there I am, maybe a number of years later, maybe at the age of 13 or 14, 13, 14, on a church retreat out in a wooded area where myself and some of the other guys who were in our Scout troop decided to explore the area and we found a lake that was frozen over and I chose to go out on that lake with a stick to test it to see if it was frozen for the rest of the guys as they went around the perimeter and got past the middle point and tap tap step and tap tap step, and fell through and they couldn't come out to help me out and I'm underneath the surface of the ice and came back to the surface, but they couldn't see that. They only saw the hole and no more Eddie kind of thing, and trying to lay out on the ice to get out of the hole thinking everything in my whole mentality to save myself, and made it, feeling kind of stupid after breaking the ice all the way in to get to the other side where they all were and everything that followed thereafter, so I overcame that fear of water. As they said, "He was so determined that he was going to make the mile swim that he dog-paddled his way around for a mile," so from that sense, I didn't have any fears about coming to the Academy and the stringent swimming requirements.

As most African Americans will tell you, I think, that have attended the Academy here, that the swimming requirements are such that they find themselves forming a subculture in the sub-pool where they have to go for remedial swims which is not the main pool. It's over in another part that; "Here's another one of us" kind of in their little subgroup. I wasn't a part of that though here, but in coming to the Academy the swimming aspects and what I'd come through in those two water scenarios there were something that I had overcome and choosing the Academy primarily for the sake of knowing that my parents were not going to be able to afford a college education vis-a-vis through any normal means, and I didn't want to risk the chance of getting not a full scholarship to some place else and that I felt I had the chutzpah, if you would, to come to the Academy and succeed, knowing that I wasn't going to quit, which became a very driving point because the processes of coming to the Academy were such that academically I was fine to come into the Academy. SAT scores and all that were; pretty wiz kind of kid.

But it was in December of 1969 and, of course, in the midst of my senior year in high school, and going through some of the Academy stuff for over the last several months, the work-up to getting in, meeting with the naval reservist who was in charge for that area and all those kinds of things, a tremendous amount of paperwork procedures that go into all those preliminary steps, come to find out that I had a hernia which made me medically disqualified to enter the Academy on its face, that December of 1969, Christmas break, there I am in the hospital getting this hernia repaired to physically qualify for entrance into the Academy, and lo and behold, maybe in February 1970, received a letter saying that I'm physically disqualified even after having the surgery, and that took me to the Naval Academy Prep School.

So I graduated high school in June of 1970, then before July 1, still 17 years old, left to go to boot camp in Great Lakes, Illinois, in preparation for coming into the Naval Academy Prep School. That's one of the requirements, that you go to regular boot camp, because if you fail or bail out along the way, you'd be going back to regular Navy per se, so taking that added step, that extra step, where once you became regular Navy, 17 years old, getting on the plane, leaving home for the first time; very traumatic steps, and I got to boot camp and Naval Academy Prep School.

All the NAPSters there are going to regular boot camp. We're all green as green can be and certainly survived that experience and went on to Naval Academy Prep School where I found that most of the people who were there were people who were academically challenged in getting into the Academy. That I was kind of in a unique class of individuals where I had the academics already. NAPS made itself out to be almost like a academic boosting school when I wasn't there for that purpose.

Schneller: So you coasted?

Graves: NAPS was nothing [laughter]. It was absolutely nothing. It felt like a year taken away from me in a sense because the disqualification was that I had had surgery within six months prior to entrance into the Academy, so-

Schneller: Why did you decide to go through with it instead of trying for a scholarship at some other school? Was the economic—

Graves: The economic factor was the overriding factor. Dad, a steel worker, at that point, and he even took on a second job as a hospital orderly. In fact, driving from Youngstown, Ohio, across the Pennsylvania border to Newcastle, Pennsylvania, to work in a hospital as a second job in addition to working as a steel worker there in the Youngstown steel mills.

You asked about the racial climate in Youngstown and what'd your parents taught you about race in that sense. You know, we grew up, and Youngstown has a nickname called Little Chicago where; and it's even still that way today. Here we are the year 2000, we're talking like 40 years ago, but Little Chicago in the sense that that area there is heavily controlled by the Mafia and there were a significant number of car bombings going on at that time period, so we were desensitized quite a bit from; and that to me wasn't even race though. It wasn't like these white mob guys are trying to disenfranchise the black person. That really wasn't what was happening in my mentality.

My mentality was is that these white mob guys are trying to keep in line their own troops because the people being blown up are like other white people, people pissing them off in whatever fashion, so you'd read about a car bombing here and a car bombing there and okay, I know there was another car bombing last night. Pretty much standard fare, so I didn't grow up with any what I call racial disrespect for white people, that they've always had an even keel in my life and in my upbringing. That's why on initial face it wasn't difficult from a race perspective coming to an institution that was traditionally white per se.

Schneller: Do you know what it was that first sparked your interest in the Naval Academy? Was there a recruiter visiting your high school? Or had had you some; well, given what you said about swimming, I don't know that you would have had some longing for the sea to become an officer or something like that.

Graves: I think probably most of it than anything else that I think I was developing my life plan at that point and that I think I felt that the best way to; this might sound strange. The best way to do what I wanted to do, to take complete control of my life for what I was trying to do at that time period, I had this longing for southern California, this longing for the ocean, and boy, I was dead set on achieving that, and as it was, I, ooh, boy, if I go to the Naval Academy, graduate from there, get a duty station in San

Diego, I'd have achieved what I was trying to do, so somewhere long ago, kind of like from a strategic planning kind of thing, I had mapped out that I wanted to have the experience of southern California.

Schneller: So you did think about a naval career beyond the Academy? It wasn't just going to the Academy for an education and then see what happens.

Graves: I'm not so sure as much as a naval career, no. I was thinking about a very selfish career of getting out to California in some fashion or another.

Schneller: Not necessarily in uniform?

Graves: Oh, absolutely not. In fact, I envisioned myself as a surfer and went and achieved that. Became an African American surfer back there in the late '70s.

Schneller: How did your family and friends respond to your decision to join the Naval Academy and how was the Navy perceived in your community? I guess we'll start with how did your family and friends respond to your decision to go?

Graves: They were absolutely flabbergasted with support in the sense that they; they probably recognized the significance of it more than I at that point. You know, I've got some of the old newspaper clippings from the announcements of going to prep school and the Naval Academy and all that, and my goodness, when you take a look at those, my picture's in the paper back then, that that's the response that was from the community, that this was something that was newsworthy and a black boy from the east side going off to the Naval Academy Preparatory School for subsequent entrance into the Naval Academy, and I probably haven't looked at those in some time but that's kind of how that, you know, the swelling of support of in a sense maybe almost like people living vicariously through me, that they were more proud of the events taking place. I'm just kind of doing things, you know, not trying to "map a legacy" or any of that at that point because you certainly don't think about those kinds of things at that age when you're 17, when you're hard charging ahead and you're memorizing texts and speeches and those kinds of things for the debate competitions and the oratorical contests per se, that, my goodness, you already have enough on your plate as to; if you had any foresight as to that this in 30, 40 years from now was going to be meaningful, I would think that you would be hedging at that point, that that wouldn't really be genuine. Those kinds of thoughts don't enter your head. You just continue to do what you're doing and to provide the degree of excellence that you can for all of your accomplishments.

Digressing just a bit, I had a huge achievement when I was like in; I can't remember right now. It must have been; I was very, very young, like maybe 4th or 5th grade, and I found myself competing in the high school spelling bee.

Schneller: Wow. That's impressive.

Graves: Little me in—

Schneller: 4th and 5th grade in the high school.

Graves: I found myself competing up against the high school kids. I was a tough little son-of-a-gun because there are and is a right answer and as it relates to spelling, you articulate, spell it, say it and sit down. Tough little nut to crack there at a very early age and each time I spelled a word correctly, and I think I made it down to the final three, oh my God, it's like wow [laughter] the buzz in the crowd, I mean the intensity and the feeling of that, little smug little son of a gun, you know [laughter] God damn, that kid's bright, as I finally lost, misspelled a word kind of thing.

I felt that sense of accomplishment and I felt that sense from the standpoint of, wow, you're competing at a level that's like four or five times removed from where you are. Kind of knew it then. Yes, knew it then, and the guys and the one girl that beat me academically, they beat me, so the three people ahead of me in that high school class, you can count on any one of them, my friend, to this day today, that those people are sharp as shit because they are. James Brown, John Lee, and Jeanette Gismondi, and Boyd Ed Graves. Those three, one of them has passed, but it was fun having that kind of academic challenge in high school. Those of us at the forefront striving for all "A"s in everything that we did.

That's the kind of background that I brought here to the Academy and since I wasn't coming here as an athlete, black athlete, but coming here to enhance my academic abilities, accepting the challenge, and I think that kind of shows with putting the Mandarin Chinese into the curriculum. They go, "Duh, what a fool you are? Wasn't it tough enough, Mr. Graves there at the Academy without adding Mandarin Chinese on your curriculum?" And now you see the number of hats I'm wearing at the same time while I was here, so I took full advantage of the academic environment here to soak in as much as I possibly could.

Schneller: Was the transition from being a civilian to being a plebe difficult for you?

Graves: Well, not after having gone through the Naval Academy Preparatory School.

Schneller: Right.

Graves: You need an extra year for that to take place, if you see what I'm saying. I didn't need to bone up another year of academics.

Schneller: Once the fall came around and you went to your company, was that the 9th Company?

Graves: We initially came in at 7th Company and ended in 9th Company and the color company at that our senior year.

Schneller: How would you characterize the midshipmen's attitudes towards race?

Graves: Upperclass?

Schneller: Your classmates and then the upperclass.

Graves: Well, you know, I'm sure for many of many of my classmates, they had had very little interaction with black people, Negroes, or the name evolution that we've gone through.

Schneller: They'd seen them on TV, but—

Graves: Even in limited capacities there, some subservient capacity, I'm sure, so there's that natural inclination that we are what we see on television. That, "Oh, he's black, what, a porter, shoe shiner, janitor," kind of environment, so some of my classmates certainly had no interaction with blacks before and here they were thrust into a group with a black person and for those who were thrust into the room with a black person, transition had to take place mentally physically and emotionally.

I probably think today that, and certainly speaking bluntly, that there are a couple of my classmates who are still from an old school of thought. My company mates, just a couple, whereas on the other hand, though, there are so many more who are as devoted and loving to the brotherhood that this Academy manifests that I could pick up the phone right now and call if I were in need of something and they certainly would make an attempt to, if they didn't have it, to make a call on my behalf to try to find it, whatever it might be that I might need, so the brotherhood has lasted, my friend, over the 25 years that we've graduated from this place and the 29 years that we've been affiliated with this place that it's as thick as blood.

Schneller: So you would include white classmates among your closest friends as well as other African Americans?

Graves: My white classmates were probably closer than other African Americans because you're one in the wilderness. You have to rely on these guys. There may be other blacks here, but, again, I didn't grow up in a racially-charged environment, where I felt that those were the only people I could rely on, other blacks, per se, because I had grown up in an environment that was mixed from the standpoint of best friends being one Hungarian Catholic and the other immigrant German, the kids I played with, right there, rode our bikes with kind of thing, which, again, this is different, though, because most people, late

’50s, early ’60s, were in polarized environments to some extent.

Schneller: Certainly. De jure segregation down south and de facto up north in a lot of cities.

Graves: Yes, but our little, our little island of reality, small steel mill town, northeastern Ohio, was such that, in a sense, I guess, you could call it a kaleidoscope, I guess. It was a wonderful environment, absolutely wonderful environment. I mean, the Hungarian house next door, our house here and they’re baking Hungarian bread and cookies, and oh, the ambience, oh, you looked forward to that.

Schneller: Sure.

Graves: They had the little waffle-like pastry kind of thing that they make the cracker-kind of cookie thing—

Schneller: Oh yes, pizelles.

Graves: Oh, my goodness.

Schneller: Did you have pierogies, too?

Graves: Oh, absolutely. That was the kind of environment that I came from and our homes were such that we initially had a plum tree, grapevine. They had a cherry tree, apple tree, and so, boy, we’d just—

Schneller: Trade fruit.

Graves: Yes, it was like; it was really a comfortable growing environment, and I didn’t think life got any better than that, to go out and pick off a plum when they’re ripe and the Bing cherries that we had there. Oh, you'd have to wait and you know what I mean, and if you waited until they were just right, they were just so wonderful, oh, to pick off the tree, so you're taking me back a few years, so I can almost taste them here [laughter] to that extent.

Schneller: How would you characterize the upperclassmen's attitudes towards race? Your squad leader and then other upperclassmen in your company, in particular?

Graves: Well, [laughter] you must know we're talking Naval Academy back then.

Schneller: Right.

Graves: We're talking Naval Academy 1971. What was it like to be a black plebe at an all-white Naval Academy before affirmative action? That's the question you're asking.

Schneller: Correct.

Graves: Well, what would you think?

Schneller: Well, I would imagine that it depended on how the people in your company felt about race.

Graves: So. Put it in perspective. You're at an all-white school.

Schneller: The chances are—

Graves: There's no such thing as affirmative action.

Schneller: Right.

Graves: You've got a—I'll be blunt—you've got a nigger in your company.

Schneller: Well, the pattern that I've seen is that you would have received a lot of extra attention and—

Graves: Of a negative fashion.

Schneller: Right.

Graves: Absolutely, and that's exactly what happened.

Schneller: Now, was it as far as an effort to drive you out?

Graves: You better believe it. Every God damn day. Every God damn day. In fact, it got so bad that they had to have a meeting.

Schneller: What kinds of things were done?

Graves: Well, very derogatory, unfortunately, and you kind of blank out as much of that as you can because you're on a mission and you're on a purpose and very hurtful things. Of course, they're all racially motivated. John Dean, Class of '73. I was his pet project to run my black ass out of this God damn school. Every God damn session. Every meal, come-around, outside his door. Ten minutes ahead of any other plebe coming around to anybody else. Braced up. Do you know what braced up is? Yes. Braced up outside his door with all rules, regulations, rates in memory, every meal, even over the weekend, to the standpoint that secretly they were laughing because I was his lawn jockey.

Schneller: What do you mean by lawn jockey?

Graves: What do you mean by lawn jockey, when you hear the term?

Schneller: A colored statue out on a lawn, that's what I—

Graves: A colored statue out by his God damn door every God damn meal. Every meal. Even over the weekends. So derogatory to the standpoint that he allowed my white classmates to walk past the lawn jockey and come into his room and play the stereo.

Schneller: Now, did he tell you he wanted to run your black ass out or use other words—

Graves: Well, my friend, it came to the standpoint of having a meeting over the hazing that he was providing. It was pretty one-on-one with him dead set on running my black ass out of here. Dead set on it. Absolutely. In fact, they had to have a meeting at some point as to "What the hell are you doing to this plebe?"

Schneller: Who convened the meeting?

Graves: It was convened by other classmates of his.

Schneller: Classmates of his?

Graves: That's right.

Schneller: Were there any other black classmates of his involved?

Graves: That's right.

Schneller: Do you have any idea how this would have come to their attention? I mean, did you talk to other classmates?

Graves: Well, my goodness, when you've got a God damned lawn jockey—

Schneller: Lawn jockey—

Graves: You know, where every meal, even over the weekend, they see you in that capacity with him grilling me like a son of a gun. Couldn't run me out, though. He couldn't run me out. It's going to have to happen another way. This guy is not quitting.

Schneller: Did the treatment deprive you of food and sleep and time to study and all that?

Graves: Would it deprive you of that?

Schneller: I would think so.

Graves: Okay. Yes. The same thing happened down in the mess hall. Couldn't eat. He's always asking me fucking questions. "Get Graves down here."

TAPE 1 SIDE 2

Schneller: . . .talking about your being hazed by, was it Davis?

Graves: John Dean.

Schneller: Dean, okay, and what other kinds of things did he do? Obviously you talked about excess come arounds. Did he have you doing a lot of physical exercises? Sweating pennies to the wall?

Graves: Yes, just the excessive bracing assumingly for certain periods of time. It was just incredible, I thought.

Schneller: And how did you know that he was doing this because you were black?

Graves: I think that began to manifest itself. That certainly if it were only for a day or on alternate days, or a week or on alternate weeks, it would be different, but it was every day, every meal.

Schneller: And you were the only one?

Graves: The only one for the longest time.

Schneller: Did some of your classmates start to avoid you, so as not to become targets themselves?

Graves: No, they couldn't do anything. They were powerless. My classmates were powerless and other people were taking shit, but they're trying to save their own ass.

Schneller: Right, and then in some cases, people in order to do that will avoid associating with somebody who's obviously a target.

Graves: Well, yes, but you're a target by definition, you know. You're black. You stand out. You've got an all-white company and there's one Hispanic guy, but you stand out, and so, "God, I don't want to be around the black guy. He's a target," by definition.

Schneller: Yes. So, in other words, there was a certain amount of avoidance of you on the part of your classmates.

Graves: You know, some of my company mates never lost faith and trust and hope in the system and have always been there for me. Still are today.

Schneller: Okay. You mentioned off the tape about a meeting that was held to stop your harassment. When did this meeting occur and how did—

Graves: Not soon enough [laughter]. And not that the harassment ended either, but I forget the time frame when that occurred. It had; this stuff had gone on for so long that I think it occurred after I had an explosion. After I didn't want to take it any more.

Schneller: Would you like me to shut this off? [end of recording]

PHASE III- B

PRESS

www.boydgraves.com

"Graves has an encyclopedic mind. He can pull numbers out of the air from reports he read 20 years ago. . ."

POZ Magazine

"The Secret Plot To Kill African Americans" December 2000

PRESS RELEASE: FOR IMMEDIATE RELEASE

JANUARY 15, 1999

FOR MORE INFORMATION CONTACT:

BOYD E. GRAVES
800.257.9387
e-mail: ed@boydgraves.com
www.boydgraves.com

AIDS BIOENGINEERING ON TRIAL

Cleveland, OH. THE ISSUE OF GOVERNMENTAL INVOLVEMENT IN THE CREATION, PRODUCTION AND PROLIFERATION OF THE AIDS VIRUS WILL BE RESOLVED IN THE FEDERAL COURT IN THE NORTHERN DISTRICT OF OHIO BEFORE FEDERAL JUDGE **LESLEY BROOKS WELLS**. THE COMPLAINT (**GRAVES V. COHEN; 98 CV 2209**) FILED BY **BOYD E. GRAVES** ON SEPTEMBER 28, 1998 CONTAINS DIRECT EVIDENCE OF BIOENGINEERING, INCLUDING THE CONGRESSIONAL TESTIMONY OF THE PENTAGON GIVEN ON JUNE 9, 1969 AND JULY 1, 1969 BEFORE A U. S. HOUSE OF REPRESENTATIVES SUBCOMMITTEE. "**THE PENTAGON'S TESTIMONY OF JUNE 9, 1969 LEAVES NO REASONABLE DOUBT OF THE ABSOLUTE CULPABILITY OF THE GOVERNMENT**." REMARKED LEAD PLAINTIFF, **BOYD E. GRAVES**, TO **JOHN MANGELS**, MEDICAL REPORTER FOR THE **CLEVELAND PLAIN DEALER**. "**THE OVERWHELMING AND SUBSTANTIAL CREDIBLE EVIDENCE WILL SUFFICIENTLY PROVE THE GOVERNMENT BIOENGINEERED THE AIDS VIRUS BY RECOMBINING A COW VIRUS AND A SHEEP VIRUS." "THE EVIDENCE WILL ALSO SHOW THIS RECOMBINED AGENT WAS PROLIFERATED IN THE SMALL POX AND HEPATITIS B VACCINES. THE MATHEMATICAL EPIDEMIOLOGY OF THE AIDS VIRUS** PROVES A LARGE NUMBER OF PERSONS HAD TO BE SIMULTANEOUSLY EXPOSED. **ADDITIONALLY, THE EPIDEMIOLOGICAL PROFILE OF THE EARLY AIDS VIRUS IN THE UNITED STATES MIRRORS WITH "REMARKABLE SIMILARITY" THE EPIDEMIOLOGICAL PROFILE OF THE COHORT STUDY GROUP OF THE EXPERIMENTAL HEPATITIS B VACCINE TRIALS. NO VIRUS DEBILITATES OR KILLS WITH SUCH PRECISE SELECTIVITY",** COMMENTED GRAVES. AN INITIAL RULING TO CERTIFY THE CLASS IS ANTICIPATED FROM THE JUDGE. A TAX DEDUCTIBLE FUND IS BEING ESTABLISHED THROUGH THE CLEVELAND FOUNDATION (ATTN: **TERRY A HANSEN** 216.861.3810). FOR ADDITIONAL INFORMATION CONTACT: **JOHN MANGELS**, **THE CLEVELAND PLAIN DEALER**, 216.999.4659; **SAMUEL W. BLACK**, CURATOR FOR AFRICAN AMERICAN HISTORY, THE WESTERN RESERVE HISTORICAL SOCIETY, 216.721.5722; OR **RONALD G. RODY**, EXECUTIVE DIRECTOR, NEW HOPE ALTERNATIVE THERAPY RESEARCH, 216.363.5060.

# # #

FOR IMMEDIATE RELEASE

February 18, 1999

FOR MORE INFORMATION CONTACT: **BOYD E. GRAVES**

800-257-9387

e-mail: ed@boydgraves.com

www.boydgraves.com/press

PENTAGON CONFIRMS THE TESTIMONY OF DR. DONALD MACARTHUR

Cleveland, OH. "TODAY I HAVE RECEIVED A FEBRUARY 1999 LETTER FROM A.H. PASSARELLA, A DIRECTOR FOR THE DEPARTMENT OF DEFENSE." "MR. PASSARELLA CONFIRMS AIDS **IS** A SYNTHETIC BIOLOGICAL AGENT". "I WILL NOW SEEK TO MOVE FOR RECONSIDERATION BEFORE THE FEDERAL JUDGE", GRAVES COMMENTED. "WE ARE ALL HOPEFUL JUDGE WELLS WILL MOVE ON HER OWN INITIATIVE AND RESTORE THIS MATTER TO THE DOCKET OF THE PEOPLE'S BUSINESS. THE PENTAGON'S LETTER COUPLED WITH THE FEDERAL PROGRAM (THE "SPECIAL VIRUS") LEAVES LITTLE DOUBT AS TO THE SUBSTANTIAL CREDIBLE INFORMATION PROVING THE ACTUALIZATION OF THE GOVERNMENT'S SYNTHETIC VIRUS". "IT IS COMMONLY BELIEVED", SAID GRAVES, "THE PENTAGON'S TESTIMONY ON JUNE 9 AND JULY 1, 1969, WAS A "COURTESY CALL" TO CONGRESS OF A PROGRAM THAT HAD BEEN LONG UNDERWAY". A REVIEW OF THE 1961 CONGRESSIONAL RECORD PROVIDES THE FRAMEWORK OF THE NECESSITY FOR STUDY OF OFFENSIVE "BIOLOGICAL " AGENTS. HOWEVER UPON CLOSER ANALYSIS, THE POLICY OF DEPOPULATION MAY BE MORE DEEPLY-ROOTED IN THE TWISTED MINDSET OF GENOCIDE; BY ANY MEANS NECESSARY. "HOMOSEXUALS AND BLACKS (AND MANY OTHERS) ARE ALL INNOCENT VICTIMS OF AN INTENTIONAL FEDERAL POLICY OF DEPOPULATION OF MINORITIES". " EVEN A CURSORY REVIEW OF NATIONAL DEFENSE SECURITY MEMORANDUM #314 (NSDM 314) WILL LEAVE YOU ON THE FLOOR, SPEECHLESS". HE SAID. "CONTRARY TO JUDGE WELLS' OPINION, THIS IS NOT A FRIVOLOUS COMPLAINT. HOWEVER, IT MUST BE CONSTRUED JUDGE WELLS' DISMISSAL OF THIS CASE (WITHOUT AN ANSWER FROM THE GOVERNMENT) IS FULFILLMENT OF HER ROLE AS CO-PARTICIPANT IN THE POLICY OF DEPOPULATION OF BLACKS AND GAYS. THE FAILURE OF THE NEW YORK TIMES TO ADDRESS THE FACTUAL CONTRADICTIONS WITH RESPECT TO THEIR NEWSPAPER IS FURTHER PROOF OF THE WELL-ORCHESTRATED **WALTZING BEDFELLOWS** OF THE ULTIMATE HOLOCAUST". "APOLOGIES AND REPARATIONS ARE MAINSTAYS IN THE FABRIC AND BELIEF SYSTEMS OF TRUE AMERICANS". "AIDS BIOENGINEERING IS NOT CONSPIRACY, IT IS PATHOLOGICAL POLICY".

# # #

PRESS RELEASE: FOR IMMEDIATE RELEASE

February 20, 1999

FOR MORE INFORMATION CONTACT: **BOYD E. GRAVES**
800-257-9387
e-mail: ed@boydgraves.com
www.boydgraves.com/press

WORLD EXPERTS DENOUNCE 'CHIMP' HIV ORIGIN

Cleveland, OH.—SCIENTISTS AND PROFESSIONALS FROM THE UNITED STATES AND AROUND THE WORLD CONTINUE TO ISSUE "UNCANNY" DENOUNCEMENT OF THE RECENT CHIMPANZEE ORIGIN OF HIV. "IN LIGHT OF THE OVERWHELMING EVIDENCE TO THE CONTRARY, IT IS NOT SURPRISING!" REMARKED BOYD GRAVES, LEAD PLAINTIFF FOR U.S. AIDS REPARATIONS. "THE HIV ENZYME IS CLEARLY A "NEW" TOOL OF DEPOPULATION". HE SAID. "ONE THAT WAS "SOUGHT", "CREATED", "PRODUCED" AND "PROLIFERATED" UNDER THE FEDERAL **"SPECIAL VIRUS" PROGRAM**.

THE NOTION THAT A PRIMATE FIRST GAVE THE VIRUS TO A HUMAN HAS BEEN COMPLETELY DISCREDITED BY THE SCIENTIFIC AND MEDICAL COMMUNITIES, BOTH HERE AND ABROAD. INCLUDED IN THE DENOUNCEMENTS ARE POSITION PAPERS FROM DRS. ALAN CANTWELL AND LEN HOROWITZ, AND PROFESSOR FRANCIS BOYLE. PROFESSOR BOYLE IS THE AUTHOR OF THE UNITED STATES BIOLOGICAL ANTI-TERRORISM ACT OF 1989. JOINING THE CRITICALLY ACCLAIMED CHORUS ARE INTERNATIONAL EXPERTS FROM CANADA, GERMANY AND SWEDEN. "I CAN ONLY HOPE THE JUDICIAL COUNCIL OF THE SIXTH CIRCUIT IN CINCINNATI, OHIO "HEARS THE MUSIC".

GRAVES AIDS BIOENGINEERING LAWSUIT AGAINST THE GOVERNMENT WAS DISMISSED **WITHOUT CONSIDERATION** BY FEDERAL JUDGE LESLEY BROOKS WELLS. "I WILL PROVIDE THE JUDICIAL COUNCIL A COPY OF THE PENTAGON'S RECENT LETTER AND THE POSITION PAPERS OF THE EXPERTS". HE SAID. "I AM SURE THE PUBLIC INTEGRITY OF THE AMERICAN PEOPLE WILL CONFER SUPPORT FOR A CONGRESSIONAL HEARING". "THERE IS CLEAR, COMPELLING AND CONVINCING EVIDENCE OF CRIMES AGAINST HUMANITY".

# # #

FOR IMMEDIATE RELEASE

February 25, 1999

FOR MORE INFORMATION CONTACT: **BOYD E. GRAVES**
800-257-9387
e-mail: ed@boydgraves.com
www.boydgraves.com/press

THE CDC CALLS AND HANGS UP ON BOYD GRAVES

Cleveland, OH— FOR IMMEDIATE RELEASE— ON 24 FEBRUARY 99 THE CENTER FOR DISEASE CONTROL ("CDC") SPOKESMAN, DR. TIM DONDERO, PHONED LEAD PLAINTIFF FOR **AMERICAN AIDS REPARATIONS**, BOYD GRAVES. "THIS IS NOT THE FIRST TIME THE CDC HAS CALLED," SAID GRAVES. **"THEY KNOW GALLO LIED TO CBS' "60 MINUTES"** AND TO THE PEOPLE OF THE UNITED STATES AND INDEED, THE PEOPLE OF THE WORLD".

CDC SPOKESMAN DONDERO STATED, "THIS MIGHT JUST HELP YOU A LITTLE BIT, GALLO DIDN'T DISCOVER AIDS (AS HE ANNOUNCED FOR THE UNITED STATES AND TO THE WORLD IN 1984)". GRAVES REPLIED: "WE ARE *WELL BEYOND* THE "STATE-SANCTIONED PROPAGANDA" OF DR GALLO IN 1984. WE ARE MORE CONCERNED WITH THE "STATE-SANCTIONED EXPERIMENTS" GALLO WAS ENCOURAGED TO CONDUCT, IN HIS POSITION AS PROJECT OFFICER FOR THE FEDERAL "SPECIAL VIRUS" PROGRAM". "GALLO'S EARLY WORK REVEALS HE ISOLATED THE HIV ENZYME AS EARLY AS 1967. HIS FIRST PAPER ON AIDS IS FIFTEEN YEARS BEFORE THE ANNOUNCEMENT OF THE VIRUS"! GRAVES SAID. "I SUGGESTED TO DR. DONDERO THAT 'HE GET A COPY OF **PROGRESS REPORT #8, AUGUST, 1971; the Special Virus program, sponsored by the United States of America; Dr. Robert Manaker, chairman. Dr. Gallo's project leadership is cited on page 104, and his experiments are listed on page 335.** "IT IS VERY CLEAR DR. GALLO *PURPOSEFULLY CONCEALED* THE "SPECIAL VIRUS" PROGRAM AND HIS INVOLVEMENT." "DR. GALLO DOES NOT MENTION THE "SPECIAL VIRUS" PROGRAM IN HIS BOOK, "VIRUS HUNTING", NOR DID HE MENTION THE "SPECIAL VIRUS" ON HIS **CBS' "60 MINUTES"** APPEARANCE.. DR GALLO BELIEVES AIDS COMES FROM MONKEYS BECAUSE "A JOURNALIST TOLD HIM SO" AND, "THAT HE ONLY BELIEVE WHAT HE READS IN THE NEWSPAPERS" (see "Virus Hunting", page 227 and **New York Native,** 9/9/84. HOW DOES THE WORLD IGNORE THE 1984 CONCLUSION OF THE AMERICAN ASSOCIATION FOR THE ADVANCEMENT OF SCIENCE ("AAAS")? GRAVES ASKED. "THE AAAS CONCLUDED THAT AIDS WAS THE PRODUCT OF ***HUMAN MANIPULATION OF GENES".*** DR. DANDERO FROM THE CDC ABRUPTLY HUNG UP THE PHONE.

"DR.GALLO DOES NOT MENTION THE "SPECIAL VIRUS" BECAUSE THE PROJECT IS SECRET AND SELECTIVE AND IN ACCORDANCE WITH POLICY PLANNING STUDY #23, **FEBRUARY, 1948,** BY GEORGE MCKENNIN OF THE UNITED STATES GOVERNMENT". "THE MILITARY'S DEVELOPMENT OF AN OFFENSIVE BIOLOGICAL WEAPON IS CONFIRMED IN THE FEBRUARY 9, 1999 LETTER FROM DEPARTMENT OF DEFENSE DIRECTORATE, **A.H. PASSARELLA.** HOWEVER A REVIEW OF THE **JUNE 9, 1969** TESTIMONY IS ESSENTIAL TO PROOF OF "ABSOLUTE EVIDENCE" (EVIDENCE BEYOND ANY DOUBT) OF THE "COORDINATED PLAN" TO "SELECTIVELY DEPOPULATE", ALTHOUGH THE MCKENNIN STUDY IS OVER 50 YEARS OLD"!

"OUR NATIONAL SECURITY IS CHALLENGED BECAUSE THE UNITED STATES IS, ***ITSELF***, ...THE GREATEST INSTRUMENT OF REPRESSION IN THE UNITED STATES AND THE WORLD". THE UNITED STATES POLICIES AGAINST HUMAN SUBJECT EXPERIMENTATION REQUIRE AGGRESSIVE REACTION UPON EXPOSURE, BEGINNING WITH AN OFFICIAL APOLOGY.

"AS EARLY AS 1962, OUR GOVERNMENT MAINTAINED AN ACTIVE POLICY OF RELEASING INFECTED MONKEYS BACK INTO THE PRIMATE COLONY OWNED BY LITTON BIONETIC IN CENTRAL AFRICA. ".

"ON JANUARY 11, 1985, DR. GALLO INFORMED SCIENCE MAGAZINE THAT 'AIDS WAS IDENTICAL TO VISNA VIRUS'". SEE, SCIENCE, page 173, January 11, 1985. "**VISNA IS A MAN MADE VIRUS, AND SO IS AIDS". SAID GRAVES**.

FRENCH AIDS DISCOVER LUC MONTAGNIER HAS FINALLY ADMITTED WHAT GALLO REFUSES TO: "THERE IS NO NATURAL EVOLUTIONARY RELATIONSHIP BETWEEN HTLV I AND HTLV III. MONTAGNIER CONCLUDES THAT AIDS IS MAN MADE AND JOINS **UNITED NATIONS AIDS** EXECUTIVE, **PETER PIOT** IN SAYING THAT THE AIDS VIRUS TOOK MANY STEPS IN LABORATORIES AND WAS NO ACCIDENT." SEE, **April 8, 1998**, **New York Amsterdam News**.

"THE 'SCHEMATIC' OF THE "SPECIAL VIRUS" (pg. 62, ***a five page fold out***) LISTS EACH OF THE MANY STEPS AS OUTLINED BY MONTAGNIER AND PIOT". SEE, "Special Virus" Progress Report, August, 1971.

"THESE WORLD CLASS EXPERTS NOW JOIN THIS GROWING PUBLIC CALL FOR "IMMEDIATE CONGRESSIONAL INQUIRY"". "AFTER ALL, CONGRESS did FUND THE CREATION OF THE FEDERAL PROGRAM KNOWN AS **"SPECIAL OPERATION-X ("SOX")** IN 1957, AND CONGRESS DID GRANT "OPEN" APPROPRIATIONS" FOR OFFENSIVE BIOLOGICAL AGENTS, IN 1962 AND 1969".

"WE WILL INITIATE AN ON-LINE SIGNATURE DRIVE TO ALLOW FOR A "SHOW OF SUPPORT" FOR THIS NEW CLASS OF AMERICAN VICTIMS".

"HUMAN RIGHTS IS A REAL OBJECTIVE FOR REAL AMERICANS"

"IN LIGHT OF THE SIGNIFICANT GAP IN EXPLANATIONS FOR THE DEVELOPMENT OF AIDS, THE LAWSUIT FILED BY MR. MONTAGNIER SHOULD BE REVIVED AND THE AMERICAN AIDS VICTIMS SHOULD BE ALLOWED TO JOIN THAT CASE".

"IT IS STRANGE, THE MEETING OF MONTAGNIER, GALLO AND REAGAN IN 1987 IS NOT A MATTER OF PUBLIC RECORD".

"IT IS ALSO UNDISPUTED THAT MONTAGNIER AND GALLO MET SECRETLY IN CALIFORNIA".

"WHERE ARE (THE) PROGRESS REPORTS (REPORTS 1 - 7} OF THE SPECIAL VIRUS PROGRAM?"

"WOULD NOT AN INDEPENDENT REVIEW OF THE MULTI-INSTITUTIONAL SERUM REPOSITORIES **DEFINITIVELY REVEAL** THE INTRODUCTION OF THE GOVERNMENT'S 'SYNTHETIC BIOLOGICAL AGENT' INTO THE POPULATION OF THE AMERICAN PEOPLE"?

"IT IS AKIN TO ANARCHY THAT JUDGE LESLEY BROOKS WELLS HAS SUBSTITUTED HER OPINION FOR THAT OF ANY JURY'S". "THIS IS A CREDIBLE LAWSUIT NOT THE PRODUCT OF DELUSIONAL THINKING OR ANY OTHER MENTALLY CHALLENGING DISABILITY". HE SAID. "IT IS CLEAR JUDGE WELLS HAS ABUSED HER DESCRETION WITH RESPECT TO SECTION 1915(E) OF THE UNITED STATES CODE. "IT IS **"EXTREMELY INAPPROPRIATE**" FOR JUDGE WELLS TO RULE UPON GRAVES, AS IF HE WERE AN OVERLY -LITIGIOUS PENAL PRISONER.

"DOES BOYD GRAVES AND THE CLASS OF PLAINTIFFS NEED ONLY PAY THE FILING FEE FOR THIS CASE TO BE HEARD"?

# # #

FOR IMMEDIATE RELEASE
9.12.00

FOR MORE INFORMATION CONTACT:

BOYD E. GRAVES
800-257-9387
e-mail: ed@boydgraves.com
www.boydgraves.com/press

BBC INTERVIEWS GRAVES ON AIDS DISCOVERY

(Washington, DC) Boyd Ed Graves, J.D. lead plaintiff for worldwide AIDS apology, completed his first interview with the British Broadcasting Corporation (BBC) on Sunday. Graves discussed details of his research into the direct and compelling evidence for proving the laboratory birth of AIDS
with BBC Executive Producer Liz Hillman.

Originally scheduled only for Saturday, the interview ran into a second day of filming, as Graves detailed the substantial evidence supporting the allegations of his federal court case and his upcoming book, "State Origin: The Evidence of the Laboratory Birth of AIDS."

"I think the second day of filming is representative of the urgency of this topic and how effectively we convey this information," Graves said.

Liz Hillman Executive Producer for the BBC indicated the interview was "very thought provoking" and said the documentary will soon be aired on the BBC network. Hillman said the working title of the documentary for which Mr. Graves was interviewed is called "Conspiracies."

"Conspiracies are based on theories, the evidence is based on fact" according to one supporter of Graves'. Supporters close to the case worry the urgency and worldwide importance of this issue may be trivialized by the 'conspiracy genre'.

"The flowchart takes this issue out of the realm of conspiracy and squarely and properly places it in the world of scientific fact. The flowchart coordinates over 20,000 scientific papers representing the fifteen year history of a federal virus development program initiated as early as 1962." Graves said during a phone interview Monday.

Graves' upcoming book, "State Origin:The Evidence of The Laboratory Birth of AIDS" is expected to be released this December in print and on-line. According to independent publisher Joel Bales, "'State Origin' is not only a case study of the Special Virus Program, it is the heroic story of one
man's tireless efforts, to bring these facts to light against all odds. The evidence presented in 'State Origin' will forever change the way the world addresses the AIDS issue."

While Graves' legal case is presently awaiting decision by the Sixth Circuit Appeals Court, he continues rallying grassroot support both on the street and in the halls of Congress. Graves federal case continues to gain international and domestic media interest; and as the pressure for independent Congressional review gains momentum, some of those who have reviewed the evidence - are now suggesting reviews in the highest courts of the world.

"If an earthquake has two epicenters, like the AIDS virus, rest assured; just as it was in the Wizard of Oz, there is someone behind a curtain!" Graves said.

###

FOR IMMEDIATE RELEASE
11.09.00

FOR MORE INFORMATION CONTACT:

BOYD E. GRAVES
800-257-9387
e-mail: ed@boydgraves.com
www.boydgraves.com/press

SIXTH CIRCUIT ISSUES 'ELECTION-DAY' ORDER AFFIRMING AIDS BIOENGINEERING AS "FRIVOLOUS"

Cleveland, OH.—The civil complaint filed against the United States for the creation of AIDS has been dealt another 'underhanded' blow. In an order issued on November 7th, the appeals court affirmed the lower court's ruling that the issue of AIDS bioengineering is a 'frivolous' matter. The ruling on November 7th (Election Day) is suspect for a number of reasons. Many experts and legal scholars believed the court would not issue a decision of the eleven month old appeal until after the election. It comes as no surprise the court chose election day itself to render an order after an eight month delay.

In light of the judicial error identified in the order, lead plaintiff, Boyd E. Graves will file the appropriate responsive motions with the court. "We will continue to strive to draw public attention to the federal program, the "Special Virus", Graves proclaimed earlier today. "The flowchart and progress reports are direct evidence of the creation of AIDS. They are only deemed frivolous because they are being presented by a Black man. The Sixth Circuit issued its "hidden", Election day order because of the significant legal precedents the court had to suspend. The presidential election and the true origin of AIDS will remain on the American spotlight well into the next millennium. We are certain we will compel government accountability of the 15,000 gallons of AIDS made in 1977.

The AIDS virus has been 'engineered' to have an affinity toward a "blood deficiency" in people of color. The progress reports of the hidden program reveal a plethora of secret experiments on people of color. The September 28, 1998 complaint filed by Graves, is supported by credible evidence, including the Pentagon's 1969 sworn testimony of the U.S.'s efforts to create a "new" "immune-suppressing" virus.

Our Constitution, and the sanctity of our democratic process compels the involvement of good people from across the United States to join our effort in seeking a review of the Manhattan-style Project, the "Special Virus". The program's flowchart is the quintessential "missing link" in proving the true origin of this diabolical laboratory creation."

"We can peel back the 'wizard of oz-like curtain' surrounding AIDS and enter the 21st Century with one less government secret. It is within our reach". He said.

FOR A COPY OF THE FLOWCHART AND ADDITIONAL INFORMATION CONTACT:

ZYGOTE PRESS 1-888-842-6419

###

FOR IMMEDIATE RELEASE
12.4.00

CONTACT:

N.O.A.H. 1-800-257-9387

KANSAS ORGANIZATION SELECTS BOYD E. GRAVES "PERSON OF THE CENTURY"

Abilene, K.S.—Joel Bales, president of the National Organization for the Advancement of Humanity, a little known organization in Kansas, selected Boyd Ed Graves J.D. as "Person of the Century."

In 1999, Dr. Graves isolated a federal virus development flowchart, proving conclusively the existence of an AIDS program (plot). According to medical and scientific scholars, the discovery of the flowchart will ultimately be shown to be "one of the greatest document finds in the history of the world."

The August 21, 1999 unveiling of the flowchart before the international scientific & medical communities marks the definitive moment in time, "when the AIDS helpless became the AIDS hopeful." The discovery and presentation of the AIDS flowchart proves we will eventually be able to peel back the AIDS curtain and demand accountability of the greatest human holocaust in the history of the world.

Graves' resolve and relentless perseverance will forever serve as a catalyst for others to navigate a life maze or matrix in search of truth and fact.

"If an earthquake has two epicenters, like the AIDS virus, rest assured;" Dr. Graves said. "just as it was in the Wizard of OZ, there is someone behind a curtain."

###

FOR IMMEDIATE RELEASE
12.06.00

FOR MORE INFORMATION CONTACT:

BOYD E. GRAVES
800-257-9387
e-mail: ed@boydgraves.com
www.boydgraves.com/press

SECRET REPORT REVEALS U.S. SPENT $550 MILLION TO MAKE AIDS

(Cleveland, OH) Boyd E. Graves, J.D. lead investigator of the Special Virus Flow Chart has proven the United States spent $550 Million to "develop and proliferate" AIDS. Graves says the 20 year secret "Special Virus" program is the last remaining secret of the 20th century.

"We have an opportunity as we approach the dawn of the new century to set right the course of a grossly failed population stabilization program relative to the out dated public law 91-213," said Graves.

The National Organization for the Advancement of Humanity (N.O.A.H.) has been receiving calls and emails from all over the world following their recent selection of Boyd Graves as their 'Person of the Century'. In 1999, Graves stunned the international and scientific medical communities when he presented the secret program's "flowchart" at an international medical research conference.

"We believe Dr. Garth Nicolson's response to the unveiling of the diagram is a 'priceless scientific reaction'worthy of further independent scrutiny. More scientists involved in the half billion dollar genocide project should seek to provide greater leadership to redirect the social course of mankind."

Together with our scientists we can realize the immediate medical breakthroughs for people living with HIV/AIDS and other illnesses. It is worthy to note the U.S. government has not responded to Boyd Graves' call to public debate.

Now that we have found the entire budget for the secret Manhattan AIDS project, we are absolutely positive reasonable, good Americans will seek to know more about the 'last secret of the 20th Century'.

The U.S. Special Virus program is a stealth biological program representative of a century long hunt to find a "contagious cancer" that "selectively kills". We are grateful for the calls and emails although we are not able to respond to all of them.

We are all awaiting our decision from the Sixth Circuit. We will have no chance of being heard by the Supreme Court. They have already done their work for next year. We hope to have Ed's book on the shelf by Inauguration Day.

It is only a matter of time before another U.S. President apologizes for yet another 'crime against humanity'. Each and every victim of AIDS is innocent in the face of a structured federal virus development program. Who amongst you does not support review of this federal virus program for its secrets in relation to the true laboratory origin of AIDS?

###

FOR IMMEDIATE RELEASE
01.10.01

FOR MORE INFORMATION CONTACT:

BOYD E. GRAVES
800-257-9387
e-mail: ed@boydgraves.com
www.boydgraves.com/press

U.S. PENTAGON
Admiral Craig Quigley
Att: Florence 703-697-9312

GRAVES TO BRIEF PENTAGON ON 1971 AIDS FLOW CHART

(Washington, DC) Boyd Ed Graves, J.D., lead palintiff for world wide AIDS apology is scheduled to present the "Special Virus" AIDS Flow Chart to top Pentagon official Admiral Craig Quigley, January 12 at the Pentagon. The 1971 AIDS Flow Chart coordinates over 20,000 scientific papers of a secret federal virus development program named the "Special Virus." Graves has called for immediate international review of the Flow Chart and the program since his discoveries in 1999.

According to Graves, no issue is greater than the immediate deactivation of AIDS. International medical and scientific experts have joined Graves' call for immediate review and deactivation of the "Special Virus" program.

According to virus program officer Dr. Alan Rabson, deputy director of the National Cancer Institutue, the secret program was unsucessful and "was terminated in 1977 in accordance with the Zinder Report." Graves has proven the virus development program continued at least through 1978 and produced 60,000 liters of a new synthetic immunosuppressing virus.

Graves continues to challenge any government scientist to a public debate on the true purpose and result of the United States' "Special Virus" program and Flow Chart. According to Graves, the good people of the United States can no longer deny the Flow Chart's role in the "creation," "production," and "proliferation" of AIDS.

According to Graves, "The Flow Chart will be remembered as the greatest document find in the history of mankind." While the Special Virus Flow Chart represents mass death, it also represents the world's first real hope for an AIDS cure.

In a November 7, 2000 Election Day order the Sixth Circuit court ruled Graves vs. The President of the United States and the true origins of AIDS, to be "frivolous." Graves has appealed the case to force government review and deactivation of the "Special Virus" program. According to Dr. Vincent Gammil , "The government knows full well what is in the (Flow Chart) document and that it was not meant for public consumption."

In Graves' words. "We have found the well spring of the genisis of AIDS, it is us."

###

FOR IMMEDIATE RELEASE
01.16.01

FOR MORE INFORMATION CONTACT:

BOYD E. GRAVES
800-257-9387
e-mail: ed@boydgraves.com
www.boydgraves.com/press

INTERNATIONAL AIDS ORIGIN ACTIVISTS CONFER ON MLK DAY

History was made yesterday when international AIDS activists Dick Gregory and Boyd E. Graves, J.D. met over telephone. The phone conversations have led to an exchange of information between the activists and all are hopeful the two will prepare a presentation strategy of their research for the new administration. "I have apprised Mr. Gregory of the 1971 AIDS Flow Chart and faxed it to him this evening," responded lead plaintiff for AIDS apology, Boyd Ed Graves.

The 30 year old flow chart is irrefutable evidence of an 'under-reported' federal virus development program. "There is no one who doubts Mr. Gregory's resolve as to concluding the true origin of AIDS. I am certain Mr. Gregory will have a forthcoming statement upon his review of the Flow Chart and 15 Progress Reports of the Special Virus program of the United States of America," said Graves.

###

FOR IMMEDIATE RELEASE:
01.31.01

Contact:
Boyd E. Graves, J.D.
Director-AIDS CONCERNS
the Common Cause Medical Research Foundation
800-257-9387 Don Scott, President 705-670-0180
www.boydgraves.com/press

Dr. David Satcher/Dr. Eric Goosby
Surgeon General
Attn: Shelley 202-690-5560

Dr. Alan S. Rabson, Deputy Director
National Cancer Institute
Attn: Anne 301-496-1927

GRAVES MOVING U.S. TO REVIEW A "SPECIAL VIRUS" PROGRAM

(Bethesda, MD) In a phone conference this morning, the Surgeon General's office admitted no prior knowledge of the government's secret federal virus development program during an interview Wednesday. International AIDS activist, Boyd E. Graves, alerted 'America's Doctor' to a secret federal virus development program called the "Special Virus", that persisted in government laboratories from 1962 through 1978.

The program's 1971 Flow Chart depicts, in graphic detail, U.S. development of a candidate virus that would deplete the immune system. The 1971 Flow Chart is accompanied by 15 progress reports detailing every experiment conducted under the program.

"The Flow Chart allows us to pinpoint where each experiment has specific relevance in the development of a "contagious cancer" that "selectively kills", said Graves. "We believe that any review of this program should begin with the experiments that "inhibited replication" of the federal virus. Specifically, in 1971, a drug, "n-demethyl rifampicin" was demonstrated to have this effect."

According to AIDS origin researcher, Len Horowitz, "Boyd Graves is the 'quarterback' in the people's efforts for review of this mostly-secret federal virus development program. He has done an outstanding job getting the issue to this level of government."

A spokesperson for the Surgeon General's office said they expect to be in further communication with Mr. Graves.

Graves' website (www.boydgraves.com) contains some of his research and supportive comments from scientists and medical doctors from across the United States and around the world.

"We have within our reach the ability to deactivate AIDS, the Flow Chart is conclusive proof of a stealth virus development program that sought a candidate virus for large scale production. Between 1964 and 1978, the United States spent $550 million to make the "special" virus." It is Graves' position that only an immediate independent review of the program will allow for the United States to begin the process of reconciling the darkest chapter in human history.

###

FOR IMMEDIATE RELEASE
02.09.01

FOR MORE INFORMATION CONTACT:

BOYD E. GRAVES
800-257-9387
e-mail: ed@boydgraves.com
www.boydgraves.com/press

Prof. Garth Nicolson
Institute for Molecular Medicine
714-903-2900

WORLD EXPERT JOINS GRAVES CALL FOR AIDS REVIEW

(Hunington Beach, CA) Boyd E. Graves, J.D., American AIDS activist and lead investigator of the Special Virus Flow Chart confirmed plans to meet with world reknowned scientist Dr. Garth Nicolson at the Institute for Molecular Medicine, following Nicolson's personal invitation Friday. The invitation follows an intense correspondence regarding Dr. Nicolson's involvement in the mostly secret federal virus development program.

Graves filed a lawsuit against the United States government for the "creation," "production," and "proliferation" of the AIDS virus, following his discovery of the Special Virus Flow Chart. According to many experts in the medical and scientific community the government's Flow Chart proves the U.S. made AIDS.

Graves says, "every victim of the Special Virus program is innocent and entitled to immediate reparations." Nicolson concurred with Graves today stating, "Just as you, I believe HIV-1 is a laboratory creation."

Graves has called for the immediate deactivation and review of the AIDS program and is hopeful Nicolson's leadership on the issue will further encourage the support and cooperation of the United States government.

"We have within our power, the ability to deactivate AIDS and close the darkest chapter in human history," Graves said. Graves and Nicolson are expected to make a statement Monday following their meeting.

###

FOR IMMEDIATE RELEASE
02.11.01

FOR MORE INFORMATION CONTACT:

BOYD E. GRAVES
800-257-9387
e-mail: ed@boydgraves.com
www.boydgraves.com/press

Dr. Robert C. Gallo

IHV (Institute for Human Virology)

Attn: Jerome 410-706-8614

Gallo Confirms Role In Special Virus Program

(Baltimore, MD) In a spirited phone conversation this Sunday evening between **Boyd E. Graves, J.D.** and "AIDS co-discover", **Dr. Robert C. Gallo**, Dr. Gallo affirmed his participation in the federal virus program, the Special Virus. Dr. Gallo confirmed his role as a Project Officer which the reports identified. Experts around the world now believe this under reported program to be the birth place of AIDS.

Dr. Gallo was presented a personal copy of the program's "research logic" today after he informed Graves that he had never seen the Flow Chart before.

The world awaits Bob Gallo's renewed leadership, in now seeking review of the Flow Chart and 15 yearly progress reports, of the Hershey Medical Center meetings between 1962 and 1978, that spent $550 Million to make a contagious cancer that selectively kills. Our call for review of this virus development program continues to gain support around the world.

###

FOR IMMEDIATE RELEASE
02.26.01

FOR MORE INFORMATION CONTACT:

BOYD E. GRAVES
800-257-9387
e-mail: ed@boydgraves.com
www.boydgraves.com/press

Tom Pope Show:
Attn: Brant 202-783-7280

TOM POPE HOSTS AIDS ACTIVIST BOYD GRAVES

(Washington, D.C.) Veteran radio talk show host **TOM POPE** hosts AIDS researcher and civil rights activist **BOYD E. GRAVES, J.D.** Thursday March 1, 2001 when Graves will discuss the thirty year old U.S. Special Virus Flow Chart and updates on the federal case.

The Flow Chart, discovered by Graves in 1999, has been called the greatest document find of the century by doctors around the world. The thirty year old Flow Chart coordinates over 20,000 scientific papers written by now prominent AIDS doctors including AIDS "co-discoverer" Bob Gallo, Peter Duesberg, and Garth Nicolson. Graves, along with other international medical authorities have insisted that AIDS is the U.S. government's "Special Virus" since the Flow Chart surfaced in 1999. Since the discovery, Graves has personally requested each of the doctors involved to lead in the program's immediate review.

Earlier this month, Graves briefed the U.S. Surgeon General's office on the Flow Chart discovery. America's top doctors admitted no prior knowledge of the mostly-secret federal virus development program or its' Flow Chart. The tax funded "Special Virus" program is not mentioned in any medical text books.

"The Special Virus program represents 15 years of missing medical history and deserves its proper place in the book of life," Graves said. "The people of the world demand immediate review of this federal program which produced 60,000 liters of AIDS between 1976 and 1977." "The Flow Chart document provides the kind of "technical foundation" that will ultimately lead to the de-activation of this synthetic biological agent."

Experts from around the world have posted comments on Graves' research and evidence on his web directory. It is worthy to note that top world scientist, Dr. Garth Nicolson confirmed Graves' central thesis that "AIDS is a laboratory-borne pandemic."

Graves judicial activism is focused on the immediate review of the Special Virus program for the ultimate deactivation of AIDS. Graves nine year mission to end AIDS has concentrated on the scientific etiology of the disease as an epidemic. Graves research concludes the government's special virus was designed to cull the Black population and legalized by federal policy and U.S. law in accordance with Nixon's P.L.91-213 (March 16, 1970).

"Within 66 years, AIDS will annihilate all Blacks in Africa," Graves said. "We must work together using the Flow Chart to deactivate AIDS before it is too late for the Black race. We now have the knowledge and ability to successfully deactivate AIDS and close the darkest chapter in human history. Be there for us so that we might be able to be there for you. We have found the wellspring of the genesis of AIDS, it is our nation state, it is us."

Graves will appear on the TOM POPE Show THURSDAY MARCH 1, 2001 at 12 NOON (EST).

###

FOR IMMEDIATE RELEASE
03.15.01

FOR MORE INFORMATION CONTACT:

BOYD E. GRAVES
800-257-9387
e-mail: ed@boydgraves.com
www.boydgraves.com/press

Navy Historical Society

Robert Schneller. 202-433-9782

NAVAL HISTORICAL CENTER TO CHRONICLE
ACADEMY LIFE OF BOYD ED GRAVES

(Washington, D.C.) The prestigious Naval Historical Center will chronicle the Academy life of 1975 Annapolis graduate, Boyd E. Graves, J.D. "We were able to do a preliminary interview with Dr. Graves when he attended the Naval Academy Minority Alumni Conference in November", commented Naval historian Bob Schneller and compiler of the massive amount of anecdotal information about early African American life at the U.S. Naval Academy. Schneller's book, **"Breaking the Color Barrier",** due to be released later this year will serve to preserve the "unique atmosphere" of Academy life primarily before affirmative action.

"What people have overlooked about Boyd Ed Graves is that he was the first African American President of the NAVY GLEE CLUB, Brigade staffer, boxer and a 'hell of a (football) manager. In fact there's a story that he asked to 'suit up' for an ARMY-NAVY game."

The vestiges of leadership diversity require us to include Dr. Graves' accomplishments, then and now. One need only enter his name on any internet search engine to fully understand the magnitude of his total contributions.

The taping will occur at the Navy Yard in Washington, D.C. next week. Dr. Graves is currently in seclusion preparing his April 11, 2001 brief to the U.S. Supreme Court. Following the filing of the brief, Dr. Graves will make a presentation of his research evidence, "Equal Justice Under Law" at 11:00 a.m. on the steps of the high court. Dr. Graves is credited with the discovery of the **1971 AIDS Flow Chart**. The Flow Chart proves the AIDS virus is the bi-product of a secret federal program. His brief is submitted on behalf of the innocent victims of this staged 'racial-cleansing' atrocity.

Schneller, R., "Breaking the Color Barrier: Racial Integration at the U.S. Naval Academy"

Graves, B. E., "State Origin: The Evidence of the Laboratory Birth of AIDS"

###

FOR IMMEDIATE RELEASE
04.19.01

FOR MORE INFORMATION CONTACT:

BOYD E. GRAVES
800-257-9387
e-mail: ed@boydgraves.com
www.boydgraves.com/press

United States Supreme Court

Clayton Higgins: 202-479-3011

GRAVES ALERTS U.S. SUPREME COURT TO AIDS BRIEF

(Washington, D.C.) Boyd E. Graves, J.D. today notified the Clerk of the United States Supreme Court that he will file the people's brief on April 11, 2001 on behalf of the 30 million innocent victims of the U.S. Special Virus program (1962-1978). The filing follows the Sixth Circuit's Election Day order ruling Graves' case and the origin of AIDS as "frivolous" November 7, 2000.

"This is the one case the high court can not turn aside," declared Graves. "We will buttress our brief with the petitions in support of our efforts from common America."

Our call for review of the 30 year old "blueprint recipe" for AIDS is just. The nation's high court will be under siege to adjudicate the significance of the Flow Chart and progress reports of a secret federal virus development program that produced 60,000 liters of AIDS (and other illnesses) by 1978.

According to a spokesperson for the Court's public affairs office, they anticipate Graves will attract a crowd on April 11 when he files his brief at 11:00a.m. and makes his statement on the Court's steps. They are also aware of the growing international interest this case has attracted from Argentina, to Croatia, to Germany, to Kenya, to Malaysia, to South Africa and Zimbabwe.

"We have met the extraordinary proof requirement to allow our extraordinary claims to be heard," Graves said. "We are confident the U.S. Constitution compels this Court the ultimate duty of blanket coverage of "equal protection". Poor people with AIDS must be entitled to Equal Justice Under Law. The issue of AIDS bioengineering is real and supported by credible and substantial U.S. medical and scientific documentation as well as the 'manipulative' policy decisions of Richard Nixon and others."

###

Boyd Graves Says A Government Flow Chart of Federal Research Programs, Experiments and Scientific Papers Ties Directly to the Development of AIDS.

By David Phinney / States News Service / April 11, 2001

WASHINGTON - A Youngstown man has filed charges with the U.S. Supreme Court alleging that the government secretly developed the AIDS virus. In his complaint, Boyd E. Graves submitted what he says is proof of a virus program started in the early 1960s.

Graves, whose last known Youngstown address was Greeley Lane on the East Side, says the $550 million program was coordinated between the Department of Defense, the National Institutes of Health and other government agencies to develop a disease to eradicate the world's black population.

Among the scientists he accuses of playing a key role in the project is Dr. Robert Gallo, an internationally acclaimed AIDS and HIV researcher who many view as the first to fully identify the AIDS virus in 1984.

Gallo, now with the Institute of Virology in Baltimore, called the charges odd and bizarre and denied knowing anything about a special virus program. "There was a special cancer virus program. There was nothing but cancer research in the whole program," he said, adding his association with it was limited.Nevertheless, Graves, who in 1992 was diagnosed with AIDS, remains adamant in his claims that a secret government plot ignited the epidemic. Graves is a 1970 graduate of East High School.

In the past 20 years, AIDS has taken an estimated 22 million lives worldwide. An additionial 36 million are living with the disease, according to the World Health Organization. Scientists continue to debate its origin.

Flow chart: Central to Graves' assertion is a government flow chart from 1971 that chronicles nine years of federal research programs, experiments and scientific papers that he ties directly to the development of AIDS.

"There is no greater proof than this blueprint document," says the onetime naval officer who is an attorney. "This is the greatest story of all time."

After learning he had AIDS, Graves began searching through government documents for information about possible cures. His relentless digging led him to the discovery of the special virus program documents. The program culminated with the federal government developing 60,000 liters of a biological agent in 1978 that would subsequently be known as AIDS, according to Graves.

He contends the Pentagon then spread the virus in Africa and the United States to cull the black population. He says a number of reports on those 15 programs still exist and provide even more detail to the production of HIV, the virus which causes AIDS.

The virus was released in Africa through smallpox vaccines and in New York City during a controlled Hepatitis B experimental vaccine, he says. Graves accuses the National Cancer Institute, the National Institutes of Health and other agencies of taking part in the Special Virus program.

Officials there say they are unable to respond to the allegations. Two government spokesmen said they know nothing about a virus program.

Frustrated: Graves said he has been frustrated trying to convince two lower federal courts to call on witnesses to explain the records he has discovered.

In September 1999, U.S. District Court in Cleveland dismissed his case. That ruling was upheld by the 6th District Appeals Court in Cincinnati in January.

Both courts ruled that the allegations are frivolous and vague. One judge called him delusional. With his Supreme Court filing, Graves expressed relief that his eight-year battle for a day in court may finally arrive.

"I felt for the first time that our government was listening," he said. "We are now holding our breath that this is going to be taken seriously. I feel as though I have crossed the finish line." The high court must decide if it will hear his case by May 23, 2001.

—

Let all things be done decently and in order

FOR IMMEDIATE RELEASE
04.27.01

FOR MORE INFORMATION CONTACT:

BOYD E. GRAVES
800-257-9387
e-mail: ed@boydgraves.com
www.boydgraves.com/press

DR. BOYD GRAVES TO PRESENT 1971 AIDS FLOW CHART TO UNITED NATIONS, CIVIL RIGHTS LEADERS

(Atlanta, GA) - Dr. Boyd E. Graves, J.D. U.S. Supreme Court Petitioner for Global AIDS Apology and discoverer of the secret 1971 AIDS Flow Chart will present an overhead lecture of the shocking medical discoveries and answer questions on his outstanding scientific research abstract, "**The Medical Etiology of AIDS**," Friday 8:30 pm May 4, 2001, at the African House of Praise here in Atlanta.

Friday's lecture precedes Saturday's presentation of the 1971 AIDS Flow Chart to the national UN World Conference Against Racism (UN WCAR) also in Atlanta. Dr. Graves will present the secret 1971 Special AIDS Virus Flow Chart he discovered in 1999 and answer questions regarding the people's U.S. Supreme Court petition for global AIDS apology filed April 11, 2001.

Dr. Graves' Supreme Court brief, case no. 00-9587, forces the United States' hand to deal directly with the evidence of the laboratory birth of AIDS. His judicial and human rights activism has led the world to an irreversible process of review of the secret U.S. Special Virus program (1962-1978); where slowly, more and more experts are conceding the program to be the birthplace of African genocide.

According to Atlanta lawyer Bill Price, "Dr. Graves has ignited a course of discussion that will irreversibly continue into the next millennium."

Dr. Graves' 1999 Flow Chart discovery continues to receive superlative acclaim from scientists and medical doctors around the world (see www.boydgraves.com/comments). Dr. Graves' ensuing book, "**State Origin: The Evidence of the Laboratory Birth of AIDS**," chronicles his legal and human rights activism and is based on the original scientific research abstract he will present at both conferences.

For additional information and to join Dr. Graves petition for immediate program review, please visit the web archives located at www.boydgraves.com .

###

PHASE IV

COURT DOCUMENTS

www.boydgraves.com

TO THE UNITED STATES SUPREME COURT:

" In this extraordinary request to be heard, we meet our extraordinary proofs requirement with the "research logic" Flow Chart of the 1962 - 1978 federal virus development program, the Special Virus. ("APPENDIX F", EXHIBIT 3). The U.S. Special Virus program spent $550 million dollars ("APPENDIX F", EXHIBIT 5) to make a virus and has never accounted for the program to the American people *until now*. "

Boyd E. Graves, J.D.

Petitioner for Global AIDS Apology,

United States Supreme Court

April 11, 2001

Docket as of February 28, 2001 11:52 pm Page 1

Proceedings include all events.
1:98cv2209 Graves v. Cohen, et al TERMED Vecchi CAT 03

TERMED Vecchi CAT 03

U.S. District Court
Northern District of Ohio (Cleveland)

CIVIL DOCKET FOR CASE #: 98-CV-2209

Graves v. Cohen, et al Filed: 09/28/98
Assigned to: Judge Lesley Brooks Wells
Demand: $0,000
Lead Docket: None
Dkt# in other court: None

Jury demand: Plaintiff
Nature of Suit: 442
Jurisdiction: US Defendant

Cause: 42:2000e Job Discrimination (Employment)

BOYD E GRAVES
plaintiff

Boyd E Graves
[COR LD NTC] [PRO SE]
1008 Elbon Road
Cleveland, OH 44121
216-382-9252

v.

WILLIAM S COHEN
defendant

THURMAN DAVIS
defendant

JOHN WODATCH
defendant

Proceedings include all events.
1:98cv2209 Graves v. Cohen, et al TERMED Vecchi CAT 03

9/28/98 -- FILING FEE: IFP (km) [Entry date 09/30/98]

9/28/98 1 MOTION by pla to proceed in forma pauperis (1 pg) (km) [Entry date 09/30/98]

9/28/98 2 COMPLAINT; jury demand (FL-103 issd) (11 pgs) (km) [Entry date 09/30/98]

9/28/98 3 CIS filed by pla. Recommended Track: Standard (1 pg) (km) [Entry date 09/30/98]

9/28/98 -- ASSIGNMENT OF MAGISTRATE JUDGE pursuant to Local Rule 3.1, Assignment of Cases. In the event of referral this case will be referred to Mag. Judge Nancy A. Vecchiarelli . 1 pg (km) [Entry date 09/30/98]

9/28/98	2	MOTION by pla to certify class action part of cmp. (km) [Entry date 09/30/98]
10/28/98	4	MEMORANDUM OF OPINION AND ORDER granting motion by pla to proceed in forma pauperis; further, pltf's claims re the transmission of the AIDS virus are dismissed purs to sec 1915(e); the Ct certifies that an appeal from this decision could not be taken in good faith; if pltf wishes to proceed on the employment discrimination he must submit w/in 30 days of the date of this order, a legally sufficient amd complt w/all relevant facts & ptys; this shall supercede the original complt; a completed USM form & 2 completed summons for each deft shall be provided by pltf; if amd complt is not filed w/in the time period set forth, this action will be dismissed. [1-1] (issued) (3 pgs) Judge Lesley B. Wells (dh)
11/2/98	5	MOTION by pltf for reconsideration of the Memo of Opinion & Order dated 10/28/98 (9 pgs) (dh) [Entry date 11/03/98]
11/18/98	6	NOTICE by pltf of change of address. (1 pg) (kv)
1/19/99	7	ORDER granting mot by pltf for reconsideration of the Memo of Opinion & Order dated 10/28/98 but adhering to previous order [5-1]. His clms will not be reinstated. Mr. Graves is given lv to file a legally sufficient amd cmp setting forth all relevant facts relating to his clm of discrimination, naming all defts agaisnt whom he wishes to proceed. If Mr. graves does not file an amd cmp on or before 2/19/99, this action may be dismissed. (issd) (3 pgs) Judge Lesley B. Wells (kv) [Entry date 01/20/99]
1/22/99	8	MOTION by pltf for recusal, & for stay of execution of the Ct's order of 1/19/99 pending reassignment (7 pgs) (dh)
1/25/99	9	ORDER denying pltf's motion for recusal [8-1], & denying motion for stay of execution of the Ct's order of 1/19/99. [8-2] (issued) (1 pg) Judge Lesley B. Wells (dh)

		[Entry date 01/27/99]
1/27/99	10	MOTION by pltf Boyd E Graves for stay of proceedings in response to Ct's order of 1/25/99 (2 pgs) (jk)
2/24/99	11	ORDER denying motion for stay of proceedings in response to Ct's order of 1/25/99 & ordering pltf to file an amd complt & appropriate forms, complying w/the ct orders of 10/28/98 & 1/19/99, or this action may be dismissed. [10-1] (issued) (2 pgs) Judge Lesley B. Wells (dh)
3/16/99	12	MOTION by pltf for appointment of counsel (17 pgs) (dh)
6/9/99	13	ORDER denying motion by pltf for appointment of counsel [12-1] (issued) (2 pgs) Judge Lesley B. Wells [EOD Date: 6/9/99] (dh)
9/16/99	14	ORDER denying as moot motion by pla to certify class action; pltf may file an amd complt re his employment

		discrimination claims by 10/15/99; pltf shall provide 1 completed USM form & 2 completed summons forms for each deft named in any amd complt he may file; if pltf fails to comply w/the above the Ct may dismiss his complt w/prej. [2-1] (issued) (3 pgs) Judge Lesley B. Wells [EOD Date: 9/17/99] (dh) [Entry date 09/17/99]
9/27/99	15	CLASS MOTION by pltf Boyd E Graves for reconsideration of order of 9/16/99 (41 pgs) (jk) [Entry date 09/28/99]
10/27/99	16	ORDER denying pltf's motion for reconsideration of order of 9/16/99; given the history of this case & Mr. Graves's failure to comply w/the Order of 9/16/99, this case is dism w/prej, & Mr. Graves is enjoined from filing any additional docmts in this action. The Clerk is to return to pltf any further docmts he submits for filing in this action. [15-1] dismissing case (issued) (3 pgs) Judge Lesley B. Wells [EOD Date: 10/28/99] (dh) [Entry date 10/28/99]
11/18/99	17	NOTICE of Appeal by pltf Boyd E Graves re: order [16-2] (cc: all counsel-notice only. USCA-notice and doc(s) mailed on 12/1/99) FEE NOT PAID (2 pgs) (shh) [Entry date 12/01/99]
12/16/99	18	MAIL Returned addressed to pltf Boyd E Graves - Notice of appeal; No forward order on file. (shh) [Entry date 12/17/99]
12/28/99	19	ACKNOWLEDGMENT from USCA for the 6th Circuit of receipt of pltf-applnt's Notice of Appeal (USCA #: 99-4476) - dt rcvd 12/3/99; dt filed 12/8/99 (1 pg) (cma) [Entry date 12/30/99]
2/15/00	--	CERTIFIED original pldgs mailed to the 6th Circuit on 2/15/00 (shh)

Docket as of February 28, 2001 11:52 pm Page 4

Proceedings include all events. TERMED
1:98cv2209 Graves v. Cohen, et al Vecchi
CAT 03

3/8/00	20	ACKNOWLEDGMENT from USCA for the 6th Circuit of receipt of certified record - dt rcvd 2/18/00; dt filed 2/21/00 (1 pg) (shh)
11/9/00	21	INFORMATION COPY of order from USCA - Affirming the dec of the District Court (USCA# 99-4476). This is not a mandate order. Clerk: L. Green; Circuit Judges: Merritt, Wellford and Siler. Dt issd 11/7/00 (3 pgs) (shh) [Entry date 11/13/00]
1/18/01	22	APPEAL ORDER: Denying pltf's motion for rehearing (USCA# 99-4476). Clerk: L. Green; Circuit Judges: Merritt, Wellford and Siler. Dt issd 1/12/01 (1 pg) (shh) [Entry date 01/19/01]
1/24/01	23	TRUE COPY of order from USCA - Affirming the dec of the District Court [17-1] (USCA# 99-4476). Clerk: L. Green; dt issd as mandate 1/22/01; Cost: None (1 pg) (shh) [Entry date 01/25/01]
2/20/01	--	RECORD on appeal returned from USCA for the 6th Circuit (shh)

[END OF DOCKET: 1:98cv2209]

UNITED STATES DISTRICT COURT
NORTHERN DISTRICT OF OHIO

BOYD E. GRAVES, CASE NO. 1:98 CV 2209

Plaintiff, JUDGE WELLS

v.

WILLIAM S. COHEN, et. al.

PLAINTIFF'S MOTION FOR RECONSIDERATION OF THE MEMORANDUM OF OPINION AND ORDER DATED OCTOBER 28, 1998

NOW COMES, plaintiff, Boyd E. Graves, who seeks reconsideration of the court's Memorandum of Opinion and Order of October 28, 1998. The background of this motion is as follows and is based on the district court's failure to properly consider three of the four exhibits attached in support of the complaint filed on September 28, 1998. Primarily, and most egregiously, the district court failed to address or properly consider exhibit one. Exhibit one consists of two pages; a congressional cover page and page 129 of United States House Bill 15090:

On or about September 28, 1998, Plaintiff, Boyd E. Graves, brought suit as the class representative against defendant, William S. Cohen, in his official capacity as Secretary of Defense, agency head for the United States Pentagon. In allegation one (paragraph one) of the complaint, plaintiff alleged (alleges) that 'Defendant, U.S. Pentagon did conspire with the United States Congress for the purposes of creating a "synthetic biological agent"'. In support of that allegation, plaintiff submitted to the district court, as exhibit one, page 129 of United States House Bill 15090, along with the cover page from the Subcommittee of the Committee on Appropriations, United States House of Representatives, Ninety-First Congress. Either by political persuasion or negative predisposition towards Americans with HIV/AIDS, the district court errs by not properly considering the sworn testimony evidence of Dr. Donald MacArthur of the Pentagon. Specifically, the district court errs by not properly considering that the term 'synthetic biological agent' was, on or about July 1, 1969, a separate category of appropriations for defendant, U.S. Pentagon. The district court's memorandum of opinion dated October 28, 1998, leads one to believe that (this deranged) plaintiff created the term 'synthetic biological agent'. Plaintiffs' complaint, if left unanswered, does provide a rational, arguable basis in fact that defendant, U. S. Pentagon did seek to create a synthetic biological agent.

The district court's Order of October 28, 1998 does not properly consider the fact that the word AIDS (acquired immune deficiency syndrome) did not exist on July 1, 1969, when Dr. MacArthur testified 'that within a period of 5 to 10 years it would be possible to be produce a synthetic biological agent, an agent that does not naturally exist and for which no natural immunity could have been acquired'. Further, Dr. MacArthur testified that as early as 1967, some research on the synthetic biological agent was slated to begin at the National Academy of Sciences-National Research Council (NAS-NRC). The district court through political persuasion or negative predisposition towards Americans with HIV/AIDS sets aside the fact that defendant, U.S. Pentagon does indeed have a prior history of proliferating and fostering a disease on an underclass segment of the American population.

With respect to this specific matter, the district court's memorandum of opinion is again in error. The district court incorrectly found that plaintiff had contracted the synthetic biological agent because of distribution to Africa of some 300,000,000 contaminated small pox vaccinations. The district court is mistaken. Plaintiff contracted the synthetic biological agent as a direct result of contaminated hepatitis B vaccinations that were initially proliferated on homosexual Americans by defendant, U.S. Pentagon. Prior to the start of defendant, U. S. Pentagon's proliferation of the synthetic biological agent within the hepatitis B vaccinations, defendant, U.S. Pentagon conspired to place its agent, Dr. Wolf Szmuness, in the position of Director of the hepatitis B vaccinations for the New York Blood Center in Manhattan, New York. As a direct result of defendant, U. S. Pentagon's agent, Dr. Szmuness, in the fall of 1978, 1083 homosexual men were vaccinated with the contaminated hepatitis B vaccine. All of the 1083 men recruited (chosen) by Dr. Szmuness were sexually promiscuous (thus their want of protection from hepatitis B), homosexual or bisexual, healthy, mostly Caucasian and under 40 years of age. Three months after Dr. Szmuness' experiment began, the first AIDS case was discovered in one of the vaccine participants. All 1083 participants died from AIDS. Beginning in March, 1980, Dr. Szmuness directed and oversaw similar vaccine experiments in San Francisco, Los Angeles, Chicago, St. Louis and Denver. In the fall of 1980, San Francisco reported to the Center for Disease Control (CDC) that a gay volunteer for the hepatitis B vaccine had come down with AIDS. A CDC report in August, 1981, of the first 26 AIDS cases revealed that all were homosexual and previously health, 20 were from Manhattan, 6 were from San Francisco and Los Angeles, 25 were Caucasian, the average age was 39 and most were well educated. According to Dr. Alan Cantwell, M.D., Los Angeles, CA, the August, 1981 CDC report essentially mirrors the epidemiologic profile of homosexuals who were injected in the hepatitis B vaccine trials. Dr. Szmuness' work does appear to be in violation of the Biological Weapons Convention Treaty of 1972, entered into force March 26, 1975, as proclaimed by Gerald Ford. President of the United States of America. Defendant, U.S. Pentagon must be compelled to release all documents, records, reports and papers relative to the diabolical infiltration of Dr. Wolf Szmuness and his subsequent work in his capacity as Director of the New York Blood Center's hepatitis B vaccine trials.

Defendant Cohen should be compelled to respond to the complaint filed on September 28, 1998.

The complaint filed on September 28, 1998 is not frivolous or irrational. The district court abused its discretion in attempting to dismiss the complaint under the non-prisoner screening mechanisms of 29 USC 1915(e)(2). In light of the foregoing presentation, it is clear the district court's determination to dismiss defendant Cohen (and other defendants) is based on the district court's political persuasion or negative predisposition to Americans with HIV/AIDS. Plaintiff's claims with respect to defendant Cohen are based in part, on the sworn congressional testimony of defendant U.S. Pentagon SEE, Exhibit 1 (resubmitted herein). Plaintiff's claims of AIDS as a biological experiment are not without precedent. Defendant, U.S. Pentagon does have an undisputed prior history of spreading a disease on an underclass segment of the American population. The district court's presentation of Neitzke and Lawler as controlling in this instant action is in error and the district court should immediately restore the initial complaint to the docket. Plaintiff does make a claim with a rational, arguable basis deeply-rooted in fact. Plaintiff's complaint clears the frivolousness bar set by section 1915(e). The standard is not whether or not the district court **believes** the plaintiff-rather, it is whether the facts are rational. Plaintiff's non prisoner claims are not based on delusional assertion. Lawler at 1199. The "screening" requirements of 28 USC 1915(e) were not intended to be used as a shield by a district court biased through political persuasion or predisposition. Upon review of Exhibit 1, reasonable persons would conclude defendant, U.S. Pentagon, did again spread a disease in the name of military research and military preparedness. In light of defendant, U.S. Pentagon's spread of the syphilis disease, the facts alleged by plaintiff can hardly be construed to be irrational or incredible. Moreover, under the most conservative construction, the original complaint does contain allegations sufficiently suggesting plaintiff, and the class of plaintiffs, have valid federal claims against defendant Cohen and defendant U.S. Pentagon for the creation, production and spread of a synthetic biological agent, which subsequently became known as Acquired Immune Deficiency Syndrome (AIDS).

The district court has jurisdiction under 5 USC 2302(b) and 29 CFR 1614.504(c).

Attached to the complaint as Exhibit 3, is a May 20, 1998 letter from the U.S. Office of Special Counsel ("SPECIAL COUNSEL"). On or about May 20, 1998, the Special Counsel chose (after a 90 day reinvestigation) to reopen a case against the U.S. Access Board for retaliation against plaintiff. On or about January 7, 1998, plaintiff formally requested EEO counseling for retaliation in accordance with 29 CFR 1614.504(c). The U.S. Access Board denied plaintiff EEO counseling and on or about February 21, 1998, plaintiff sought to have the Special Counsel reopen the prior case against defendant U.S. Access Board (MA-97-1946). Although the U.S. Access Board has refused to hire plaintiff through some five vacancies, at issue in MA-97-1946 was a specific vacancy (97-04) with the U.S. Access Board. Vacancy #97-04 was a Telecommunications Accessibility Specialist for which the agency had advertised the position twice in 1997 and found 'no applicants with appropriate skills'. Plaintiff applied for vacancy #97-04 and the U.S. Access Board sent plaintiff's application to the Office of Personnel Management (OPM) for ranking and rating. The U.S. Access Board abandoned the recruitment when OPM ranked plaintiff among the top scorers. The Access Board has not hired plaintiff through some five vacancies because he self-identified his HIV status in support of seeking a Schedule A appointment with the agency. Out of economic duress, plaintiff entered into a settlement with the agency on or about September 26, 1997, which resulted in the plaintiff receiving $26,000. However, one of the terms of the settlement required the Access Board to forward to other federal agencies (including named defendant John Wodatch) plaintiff's resume, a reference letter and his state certification of his qualification for a Schedule A federal appointment. On or about December 1, 1997, plaintiff moved from the Washington, D.C. area to Pittsburgh, Pennsylvania and contacted Mr. Eugene Nelson, Area Director for the Pittsburgh Equal Employment Opportunity Commission ("EEOC"). According to the Access Board and the settlement agreement, Mr. Nelson, as well as the other agency heads, had been individually mailed a copy of plaintiff's resume and materials on or about October 6, 1997 via a transmittal letter from Access Board Executive Director, Lawrence Roffee. On or about December 10, 1998, Mr. Nelson had not seen my resume and materials nor does his office have any mail room records of correspondence from the U.S. Access Board. At Mr. Nelson's request, **on or about December 10, 1997, plaintiff** supplied to the Pittsburgh EEOC, a copy of the October 6, 1997 transmittal letter from the Access Board's Executive Director. Out of further discrimination/retaliation directed towards plaintiff, the Access Board did not forward to the Pittsburgh EEOC, plaintiff's employment materials or the transmittal letter from the agency's executive director. The Access Board did not hire plaintiff, through some five vacancies because, in part, he is HIV Positive. In its attempts to not hire plaintiff, the Access Board hired individuals with less education and experience, all under the age of forty. In addition to plaintiff's identified jurisdictional claims, plaintiff can also establish a claim based on age discrimination. By not forwarding plaintiff's resume and materials in accordance with the settlement agreement, the Access Board proves they did not want plaintiff working for any federal disability agency. Upon learning in early December, 1997, that one recipient (Mr. Nelson) had not received the transmittal letter from the Access Board, plaintiff then sent to each of the other federal officials, a copy of the Access Board's October 6, 1997 letter and materials. Plaintiff sent to named defendant, John Wodatch, a copy of the Access Board's transmittal letter and materials. At that time defendant, Wodatch had four vacancies in his department. As it was, with officials at the Access Board, Wodatch personally knew plaintiff (and his HIV status). Defendant Wodatch did not respond to plaintiff until February, 1998. Defendant Wodatch had filled the four positions and alerted plaintiff that he had no vacancies.

Exhibit 4 of the complaint establishes a basis for employment/retaliation discrimination.

In brief, on or about December 10, 1997, named defendant, John Wodatch, Department head, Disability Rights Section of the Department of Justice received from plaintiff a copy of the October 6, 1997 transmittal letter from Lawrence Roffee, Executive Director. During the period of October 1997 through January, 1998, defendant Wodatch sought to fill four vacancies in his department of which plaintiff has specialized experience, was entitled to consideration for employment under Schedule A of the federal hiring program and was entitled to a five point veteran's preference in hiring with respect to his military service (1970-1977). Defendant Wodatch did not respond to plaintiff's December, 1997 mailing until February, 1998. Defendant Wodatch filled all four vacancies in January, 1998 and responded in February, 1998, to plaintiff that he had no vacancies. Plaintiff was not considered for federal employment by defendant Wodatch (or Access Board) although plaintiff is a graduate of Annapolis, holds a law degree and has a plethora of experience in the Americans with Disabilities Act (ADA) compliance and enforcement areas, including specialized experience as an 800 telephone number ADA 'answer man', the very same position(s) defendant Wodatch (and defendant Access Board) was seeking to fill.

Plaintiff will be able to show that defendant Wodatch personally knew plaintiff, and that defendant Wodatch knew plaintiff was HIV Positive. Defendant Wodatch's actions of discrimination/retaliation are consistent with that of officials of the Access Board and other federal agency officials comprising the leadership of the federal disability community. Defendant Wodatch is a proper defendant in this action. In summary, plaintiff has been continually and consistently bypassed for federal employment in the disability community in violation of 42 USC 1985(2) & (3). Plaintiff also has established jurisdiction under 29 USC 791. Defendant Access Board refused to hire plaintiff through some five vacancies. During the relevant times, plaintiff was subjected to four interviews and two exams, while other non-HIV applicants were hired without interview(s) or exam(s).

WHEREFORE; plaintiff prays the court will reconsider its memorandum of opinion and Order of October 28, 1998, and return plaintiff's complaint of September 28, 1998 to the docket. Plaintiff further prays the court will acknowledge this plaintiff has no prior history of paranoid or delusive behavior. Plaintiff reasserts through this affirmative presentation that his complaint filed on September 28, 1998 was not done so for frivolous or malicious purposes. Plaintiff further prays the court will certify a class of affected Americans and appoint counsel for the class and appoint counsel for plaintiff.

Respectfully submitted,

Boyd E. Graves
Cleveland, Ohio 44121

UNITED STATES DISTRICT COURT
NORTHERN DISTRICT OF OHIO
EASTERN DIVISION

BOYD E. GRAVES, et al	)	JUDGE LESLEY WELLS
	)	
Plaintiffs,	)	CASE NO. 1:98 CV 2209
	)	
v.	)	
	)	
WILLIAM S. COHEN, et. al.,	)	
	)	
Defendants.	)	
	)	

CLASS MOTION FOR RECONSIDERATION OF THE COURT'S ORDER OF SEPTEMBER 16, 1999 BASED ON NEWLY-DISCOVERED AND NEWLY-DEVELOPED EVIDENCE

This motion for reconsideration of the court's Order of September 16, 1999, is submitted based on the enclosed, newly-discovered and newly-developed evidence of a 1962 federal program entitled "Special Virus". Progress Report #15 of the Special Virus program correlates to the year, 1978. The flowchart of this hidden federal program is attached as absolute proof of the laboratory origin of AIDS (and AIDS-like chronic illnesses). SEE, 1971 AIDS FLOWCHART, the research logic of the "Special Virus" program (EXHIBIT TWELVE (12), herein). Defendant Cohen, et al (e.g., Robert Gallo, Robert Manaker, Paul Kotin, Fred Rapp, Carlton Gajdusek, Peter Duesberg) should be bound by the U.S. Constitution, specifically the First and Fifth Amendments and required to ANSWER the complaint filed in this matter. The legislative history of Congress and the intent and will of the American people require that Defendant Cohen, et al, ANSWER the specific questions and allegations with respect to class exhibit one (the sworn testimony of the U.S. Pentagon given on June 9 and July 1, 1969). Class Exhibit 1 has been subsequently confirmed by defendant U.S. Pentagon (See Exhibit 14, herein) and a top world microbiologist, Nancy Nicolson, Ph.D. SEE, transcript, " Interview with the Drs. Nicolsons", Radio Liberty, Dr. Stan Montieth, host, September, 25, 1995. Additionally, the Court must entertain the testimony of Professor Francis Boyle, Ph.D., author of the United States 1989 Anti-Biological Terrorist Attack Act. The class of plaintiffs plead with this court to rule on the evidence. The credible exhibits attached to the complaint of September 28, 1998 and the many additional documents do indeed present a legitimate right to access this court. The class exhibits are meaningful, and this class should not be precluded from their opportunity to have exposure and closure on this , the greatest human tragedy in the history of the world. Your Honor's attitude to dismiss as frivolous is reflective of how far the monkey origin of AIDS baseless propaganda has truly spun. However, the plaintiffs do not believe this court nor any other government official (Drs. Gallo, Duesberg, et al) can identify ANY epidemiological study confirming epidemic immune deficiency in primates. This court continues to conveniently overlook the fact that , 'if AIDS were ancestral, i.e., a derivative of some monkey immune virus that, "somehow" hopped species at one point; then why do we still have monkeys?

The people believe the Constitution requires that this definable group of Americans are entitled to an apology and in nearly every case, reparations. Particularly those who have some evidence of microbioinoculation.

In determining the frivolity of an issue, the court can not set aside direct evidence, as the court has done with plaintiff's EXHIBIT 1, filed on 9/28/98. The defendant's sworn testimony is direct evidence, and enactment of the 1915(e) dismissal scheme in this case, is contrary to every principle of law of the Sixth Circuit and the U.S. Supreme Court. Section 1915(e) has been enacted contrary to the intent of Congress and the legislative history of the statute and even more flagrant, the court is precluding this class from legitimate appeal! Defendant Cohen's testimony of June 9, 1969 and July 1, 1969 goes to the heart of the people's case, that there is significant, verifiable evidence of a master scheme of depopulation. The Special Virus (AIDS) program is only part of the sophisticated scheme to cull humanity, consistent with the eugenics (racist) principles of the Rockefeller Institute and Foundation. However, this court can not deny the existence of this program, with the current submission of the 1971 Flowchart (Exhibit 12). The dismissal of the named defendant pursuant to 28 USC 1915(e) is contrary to the very nature of our system of justice when one considers the overwhelming evidence in support of the class allegations. The experts have reviewed the flowchart and it is authentic. The Special Virus program isolated, stabilized and mass produced a leukemia/lymphoma virus. SEE, Progress Report #8 of the Special Virus program at 61, 104 -106, 335 (1971). Plaintiffs bring to the court's attention that HIV was originally called, "leukemia/lymphoma virus". SEE, Montagnier, et al, Science, 225-63, (1984), Montagnier, et al in "Human T-Cell Leukemia Lymphoma Virus". R.C. Gallo, M.E.Essex, L. Gross (Cold Spring Harbor Laboratory, Cold Spring Harbor, NY , 1984).

The people have definitive proof of another "non-public" human subject experiment gone awry. We have compensated our victims before, we must do so again, the Constitution requires it, the heart of true America compels it. We respectfully request that this court permit service of process upon defendant Cohen, et al.

US POPULATION POLICY

This United States scheme (that the Third and Fourth World 'rapid' population growth is a threat to the national security of the United States) is supported by U.S. population policy decisions that have been in place since 1974. It is highly probable the United States incorporated "state-sanctioned premeditated murder" into its population culling policies as early as 1967, the first year the United States started giving taxpayer money to the United Nation's depopulation wing, the Agency for International Development (AID). SEE, also, July 18, 1969, former President Richard M. Nixon's "Special Message to the Congress on Problems of Population Growth". Even as early as 1969, it was determined that "immediate steps" had to be taken to thwart the rapid population growth in the Third and Fourth Worlds. National Security Study Memorandum -200 (NSSM-200), titled, "Implications of Worldwide Population Growth for U.S. Security and Overseas Interests", dated April 24, 1974 by the National Security Council, (Henry Kissenger), best presents the (Rockefeller) illusion of the necessity for depopulation (primarily of Africa) as articulated by former President Nixon in July, 1969. The execution of NSSM-200 by President Gerald Ford as outlined in National Security Defense Memorandum #314 (NSDM#314), November 26, 1975, signed by Brent Scowcroft is equally compelling as to the eugenics mindset strangling humanity and seemingly ensuring the continuation of racism well into the twenty-first century.

However, in addition to the foregoing U.S. documents outlining and implementing the necessity to cull the human race, particularly Africans, it is National Security Council Memorandum #46 (NSCM #46) that is most dreadful with respect to the necessary premeditation of this depopulation scheme. NSCM #46 is titled, "Black Africa and the U.S. Black Movement", March 17, 1978, signed by Zbigniew Brezinski for President Carter. For the convenience of the Court I am attaching the available copy of NSCM #46 as EXHIBIT THIRTEEN(13) to this complaint. NSCM #46 is the United States's "contingency" plan to counteract African American response to the realization that the United States was depopulating Africa.

PLAINTIFF'S DOCUMENTS

The documents attached to this motion for reconsideration for service on defendant Cohen, et al are credible and require an ANSWER under the Constitution. The complaint filed on September 28, 1998 was sufficient for notice to the named defendants as to the specific allegations against them. The supportive evidence of the Constitutional right to have defendant Cohen, et al respond is incorporated in the collective exhibits of this complaint filed to date and the federal rules have been met and exceeded with respect to sufficiency of the complaint.

Additionally, the recent letters from Defendant Cohen and Senator Voinovich prove conclusively the class has presented a non-frivolous case. SEE, February 9, 1999 letter from Defendant Cohen, et al (EXHIBIT FOURTEEN (14), herein) and the August 31 and September 21, 1999 letters from Senator Voinovich (EXHIBIT FIFTEEN (15) and EXHIBIT SIXTEEN (16), herein). Senator Voinovich believes this is a credible issue. **The court must note that "frivolous" indigent complaints are rarely supported by U.S. Senators and physicians and scientists from around the world.**

Defendant Cohen's February 9, 1999 letter to plaintiff confirms plaintiff's Exhibit One. Plaintiff's Exhibit One is the July 1 (June 9), 1969 sworn testimony of Defendant Cohen, et al seeking to create an immunesuppressing contagious virus for which no natural immunity could have been acquired) SEE EXHIBIT 1, filed 9/28/98. According to Defendant Cohen's spokesperson, Dr. Donald MacArthur, on June 9, 1969, before the U.S. House of Representatives Appropriations Committee, the U.S. Pentagon concedes it was working on an offensive, contagious "synthetic" biological agent that would lead to "world-wide scourge" and "black death-type plague"

The 1971 flowchart supports Defendant Cohen's (U.S. Pentagon's)1969 sworn testimony. Further, however, the flowchart is page 61 of Progress Report #8 of the "Special Virus" program (1971). This matter should not be deemed frivolous until the Court has conducted an independent review to determine the validity and credibility of the flowchart, its research logic, as outlined, and the correlating experiments, contracts and contractors, specifically identified in the ensuing progress reports. The class believes progress reports #'s 1 - 7 of this program have been purposefully destroyed. Defendant Cohen, et al must answer this additional allegation relative to the creation, production and proliferation of AIDS, and the destruction of the progress reports. There is sufficient, additional evidence of the laboratory origin of AIDS and thus the unwitting involvement of Americans (and others) as human subject experiments. The scientific papers are replete with substantiating significance of the overwhelming proof of the bioengineering of the "Special Virus" (AIDS). Chief amongst the supportive scientific papers is the 1983 report from the National Academy of Sciences. According to the Proceedings (83:4007 - 4011), HIV and VISNA (an "ICELANDIC" sheep disease) are 'highly similar and share all structural elements, except for a small segment which is nearly identical to HTLV. This article, among others, confirms the fingerprint of man in the genetic sequencing of AIDS. VISNA is a man made virus sent to Iceland by the German government. SEE,

"AIDS: An Explosion of the Biological Timebomb?", by Robert E. Lee, (1998).
In light of the significant evidence in support of plaintiff's complaint, reasonable persons would conclude this complaint, heavily supported by substantial, credible evidence is entitled to an ANSWER from defendant Cohen, et al. Plaintiff has continually developed a ready reservoir of experts to augment the court's understandings, e.g., Dr. Cantwell, Dr. Horowitz, Dr. Nicolson, Professor Boyle, Chemist Gammill, Dr. Dorman and Don Scott. They have each publicly pronounced their conclusions as to the definitive proof of the existence of the 'hidden' Special Virus program. Inside the Special Virus is an "ultra secret" program entitled, "Manaker/Kotin-Special Virus Leukemia program ("MK-SVLP"). SEE, Progress Report #8 at pages 273 - 289. These pages of the Special Virus program now attached to the record of this matter definitively prove the iatrogenic, man to primate, medically-induced origin of the "simian" immunodeficiency virus (SIV). Id. Additionally, beginning on page 276, under "SUMMARY OF THE INOCULATION PROGRAM", the United States gives an overview of "new born" monkey inoculations and releases since 1962. Id. (Pages 273 -289 are attached as EXHIBIT SEVENTEEN (17), herein)
For any appeal in this matter, the class believes it can meet its "prima facie" burden of showing sufficient facts to further allege the Special Virus (AIDS) has a predisposition toward people of color. Progress Report #8 (1971) reveals the Special Virus program sought to determine if the Epstein Barr Virus had specific racial blood markers. The program used two test groups, one group was inner city children in Philadelphia (African American) and the other group was in Sweden. SEE Progress Report #8 at 109 and 130. Although the experiments with inner city children in Philadelphia began in 1966, the comparative study in Sweden was not initiated until April 9, 1969, the apparent strategy result of the Fort Detrick Conference held on April 4 - 5, 1969. The record is clear, the April 1969 Conference was entitled, "Entry and Control of Foreign Nucleic Acid". Defendant Cohen, et al intentionally sought to create a "new" microorganism that selectively kills based on racial "blood markers". The AIDS virus has an affinity to a deficiency of the OKT4 Epitope in people of color. SEE, "OKT4 Epitope Deficiency in Significant Proportions of the Black Population", Terence T. Casey, MD, et al, Arch Pathol Lab Med- Vol 110, August, 1986. SEE, also, "AIDS: An Explosion of the Biological Timebomb?", Lee, 1998 (glucose 6-phosphate dehydrogenase enzyme-variant human G6PD deficiency in Blacks). The results of the racial blood markers affiliated with Epstein Barr Virus were reported in the 1984 yearbook of the Stockholm International Peace Research Institute (SIPRI).

CONGRESSIONAL INTENT
Plaintiff respectfully requests the Court to reconsider the Congressional intent of Section 1915(e). The Court's Order of September 16, 1999 strongly imp[lies that the class representative submitted an unsupported, baseless complaint against defendant Cohen, et al The origin of AIDS under the auspices of the U.S. government is not a frivolous issue unworthy of a response. Plaintiff and this class of plaintiffs should not be relegated to the dismissal scheme of the poverty provision of Section 1915(e). Plaintiff is the class representative for Americans who have been injured or killed by the United States' "synthetic biological agent" (AIDS). This class has sufficient, credible evidence to support its claims of "bioengineering" (recombined animal and human virus) and "deployment" (small pox and hepatitis B vaccines). SEE, Plaintiff's Motion for Reconsideration, filed November 2, 1998. The class, in the name of the United States Constitution, is entitled to an answer to the complaint filed in this matter in that the class and the complaint meet all applicable federal rules for construction and sufficiency of complaint under the identified jurisdictions.
This Court must explain the insufficiency pursuant to Section 1915(e), of the class complaint and compelling exhibits. The Court must certify the legitimate right to appeal. The record is devoid of any prior discussion by the Court of the evidence, particularly, the June and July, 1969 sworn testi-

mony of Defendant Cohen, et al, (Dr. MacArthur)Exhibit one to the complaint, filed 9/28/98.

AMENDED COMPLAINT

Moreover, the Court must consider that Graves has amended his complaint with respect to the other named and yet-named defendants. SEE Plaintiff's Motion for Reconsideration, filed: November 2, 1998 and Plaintiff's Motion For Appointment of Counsel, filed: March 16, 1999. The docket further reflects that plaintiff has complied with the Court's request to submit service of process papers. SEE 6/23/99 service of process papers contained in the file folder of this case, signed by plaintiff.

This plaintiff has already met and exceeded the court's Order of September 16, 1999. Plaintiff's constitutional right to file suit has been precluded by the court. The court has placed an added burden (on poor plaintiffs) that, in this instant action, precludes effective, meaningful access to the court. The class right to pursue what they believe to be a non-frivolous suit has been abridged. The conduct complained of "bioengineering of AIDS pursuant to the Special Virus program" does or should "shock the conscious" of this court. SEE, Lillard v. Shelby County Board of Education, 76 F.3d 716, 724 (quoting Mertik v. Blalock 983 F.2d 1353, 1367 -1368 (6th Cir. 1993).

LEGAL ARGUMENT IN SUPPORT OF THE CLASS MOTION FOR RECONSIDERATION

The court's Order of September 16, 1999 should be reconsidered because of the substantial, credible evidence now supporting the complaint filed on September 28, 1998. However, the court's Order of September 16, 1999 should be reconsidered based on the precedence of law in the Sixth Circuit Court of Appeals and the United States Supreme Court. The court's Order of September 16, 1999 violates universally, the rights of poor people (in this case), with HIV/AIDS to equal protection and due process. The court's Order of September 16, 1999 stands in contradiction to significant, well established case law . SEE, e.g. Knop v. Johnson, 977 F.2d 996, 1003 (6th Cir. 1992), cert denied, 507 U.S. 973 (1993), Bounds v. Smith 430 U.S. 817, 823 (1977) quoting Ross v. Moffitt, 417 U.S. 600, 611 (1974) (interchanging "poor plaintiffs" for prisoners. Finally, pursuant to Procunier v. Martinez, 416 U.S. 396, 419 (1974), it further appears that the utilization of the dismissal scheme of Section 1915(e) is invalid and service of process should immediately proceed. The court must ensure that poor people (with HIV/AIDS) have adequate, effective and meaningful access. This case has persisted in the chambers of the court for an entire year. Americans (and others) with HIV and AIDS deserve better. WHEREFORE; plaintiff, Boyd E. Graves, prays the court will reconsider its Order of September 16, 1999 and allow service of process on defendant Cohen, et al. Plaintiff further prays the court will certify the class without further delay and immediately appoint counsel. Plaintiff further prays the court will affect service of process, pursuant to the summons and federal Marshall service papers filed in the docket of this matter on June 23, 1999, and deem the court's Order of September 16, 1999 moot.. These intertangled issues of the laboratory origin of AIDS and depopulation are real and worthy of jury adjudication as demanded by the United States Constitution.

Respectfully submitted,

Boyd E. Graves, lead plaintiff

Cleveland Heights, OH 44121-1429
Dated: September 27, 1999

IN THE UNITED STATES COURT OF APPEALS
FOR THE SIXTH CIRCUIT

BOYD E. GRAVES, et. al.,
Lead plaintiff-appellants,

Appeal No.: 99-4476
(Dist.Ct. #: 98CV2209)

WILLIAM S. COHEN, et. al.,
Defendants-appellees.

Pro Se Appellant Brief

APPEAL FROM THE FINAL ORDER ENTERED ON 10/27/99

(MAJOR) QUESTION ON APPEAL:

Did the district court abuse its discretion and commit other reversible errors when it "set aside" Graves' evidence in support of his allegations of AIDS bioengineering, in which to reach a finding of "frivolity" under 28 USC 1915(e)?

SUMMARY OF THE APPEAL

Between 1995 until present, plaintiff-appellant, Boyd E. Graves ("GRAVES") faces federal employment discrimination relative primarily to his disability (HIV/AIDS). During the course of his research into HIV/AIDS, Graves discovered a 1971 flowchart, part of an ultra-secret federal program entitled, "Special Virus". The "special virus" began officially in 1962 and produced 15 yearly progress reports. The archives of the National Cancer Institute houses some of the reports. The 'research logic' reveals the program was seeking to isolate, stabilize, develop and proliferate a synthetic biological agent (a "human" retrovirus). SEE Graves v. Cohen, Exhibit One, filed September 28, 1998. On September 28, 1998, Graves brought suit against the named (and yet named) federal defendants pursuant to final proceedings before the EEOC, Department of Justice and the little known, Office of Special

Counsel. On October 28, 1998, the district court dismissed Graves' allegations of AIDS bioengineering as frivolous. Graves believes the district court can not "set aside" direct evidence in which to reach a finding of frivolity under 28 USC 1915(e). The district court's final Order represents an abuse of discretion for a number of reversible reasons: 1) There is an identifiable class, 2) Graves has met the standard for appointment of counsel, 3) Graves' activities call for no curtailment or injunctive restrictions, 4) His complaint and exhibits, filed on September 28, 1998 meet and exceed the federal rules, 5) The district court erred in not allowing service of process. As a direct result of the court's action, Graves' constitutional rights and health continue to suffer, because of the excessive delay created by the district court's errant judicial activism. Graves believes this matter should be immediately returned to a neutral district court for service of process and appointment of counsel.

NATURE OF THE CASE

On October 28, 1998, the district court dismissed as frivolous, Graves' claims of AIDS bioengineering against defendant-appellee, U.S. Pentagon, et. al. In order to do so, the district court conveniently "set aside" Graves' Exhibits. Exhibit One is page 129 of U.S. House Resolution 15090, Part VI, of the Ninety-First U.S. Congress. Exhibit One is sworn Congressional testimony by the U.S. Pentagon given on June 9, 1969. The heading listed in the Congressional Record is "SYNTHETIC BIOLOGICAL AGENT". On June 9, 1969, the U.S. Pentagon informed the U.S. Congress of it's involvement in the development of the "Special Virus". In consideration of the credible history of the "special virus" program, it is reasonable to believe the program was well underway prior to 1969. This fact is thoroughly

supported by the record of the program. SEE Progress Report #8 at 2, (1971). Under the leadership of (yet-named) defendant, Robert C. Gallo, a project officer for the program, the "special virus" isolated a "human" retrovirus and co-mingled it with animal viruses.[1] Graves believes the district court is not free to set aside his evidence of AIDS bioengineering, nor enjoin him from further filings. Graves has sufficiently demonstrated that his claims of AIDS bioengineering are not frivolous and are worthy of an ANSWER from the United States.

With regard to Graves' motion for certification of the class, in his capacity as lead plaintiff, he sincerely believes there exists a live controversy worthy of further adjudication.

THE COMPLAINT FILED ON SEPTEMBER 28, 1998 IS SUFFICIENT FOR SERVICE OF PROCESS

Graves' complaint meets and exceeds the federal requirements for sufficiency under Fed.R.Civ. P. 8(a). Additionally, Graves' motion for reconsideration (amended complaint) filed on November 2, 1998, clearly cures every (if any) defect identified by the district court on October 28, 1998. The sworn Congressional record, the flowchart and progress reports of the ultra secret program, and the substantial, credible scientific evidence require an ANSWER, consistent with every other legitimate demand of the U.S. Constitution. Equally, the October 13, 1999 press release of Dr. Len Horowitz (Appeal Exhibit "A", herein) identifies the Chairman of the National Security Advisory Board (Colonel Jack Kingston) as a significant professional objection to the district court's determination.[2] Graves believes his

---

[1] The record of the co-mingling of human and animal viruses is located at pages 273 –289 of Progress Report #8 (1971). Any cursory review of these specific experiments will conclude the United States co-mingled human/monkey and sheep viruses, inter alia, in the development and production of the "special virus".

[2] Similar objections have been raised by Dr. Alan Cantwell and Professor Francis Boyle, author of the 1989 U.S. Anti-Biological Terrorist Attack Act.

timely filed motion for reconsideration on 9/27/99, best exemplifies the totality of the substantial evidence against the United States.

CONCLUSION

The district court has abused its inherent powers and has 'actively' sought to thwart or preclude this appellant from well established Constitutional rights of due process and equal access, inter alia. The appellant and the class are both entitled to service of process and an ANSWER. Graves believes the United States should be compelled to ANSWER the credible claims of AIDS bioengineering. The judicial activism exhibited by the district court is akin to the current "wall of silence" permeating the medical and scientific communities. The 1971 flowchart of this grotesque federal program is the indisputable "missing link" in the etiology of AIDS. The people can now 'easily' duplicate the program's experiments. The people can now [illegible]ely the AIDS virus is a chimera. As the attached letter from Senators DeWine and Voinovich indicates (Appeal Exhibit "B", herein), the legislative branch of our government is indeed spineless with regard to any investigation of the "Special Virus". Perhaps it is because NONE of them have AIDS. Please return this matter immediately to a "fair-minded" district court. The people have a Constitutional right to accountability for the appalling state conduct of Dr. Gallo, Dr. Carlton Gajdusek, Dr. Robert Manaker, Dr. Paul Kotin, et. al.

Respectfully submitted,

Boyd E. Graves, pro se

4

Exhibit 1

U.S. Department of Justice

United States Attorney
Northern District of Ohio

1800 Bank One Center
600 Superior Avenue, East
Cleveland, Ohio 44114-2654

January 20, 2000

Mr. Boyd E. Graves
2700 Washington Street
Cleveland, Ohio 44113

Re: *Boyd E. Graves v. The President of the United States, et al.*, Court of Appeals Case No. 99-4476

Dear Mr. Graves:

Enclosed is a copy of the Appearance of Counsel form which was mailed to the court on January 19, 2000.

Very truly yours,

Lisa Hammond Johnson/pdh

Lisa Hammond Johnson
Assistant U.S. Attorney
216/622-3679

Enclosure

pdh

UNITED STATES COURT OF APPEALS
FOR THE SIXTH CIRCUIT
100 EAST FIFTH STREET, ROOM 532
POTTER STEWART U.S. COURTHOUSE
CINCINNATI, OHIO 45202-3988

LEONARD GREEN
CLERK

CAROL FIELD
(513) 564-7024

Filed: November 7, 2000

Boyd E. Graves
3844 E. 140th Street
Cleveland, OH 44128

Lisa H. Johnson
Office of the U.S. Attorney
600 Superior Avenue, E.
Suite 1800 Bank One Center
Cleveland, OH 44114-2600

RE: 99-4476
Graves vs. Cohen
District Court No. 98-02209

Enclosed is a copy of an order which was entered today in the above-styled case.

(Ms.) Carol Field
Case Manager

Enclosure

cc:
Honorable Lesley Brooks Wells
Ms. Geri M. Smith

No. 99-4476

UNITED STATES COURT OF APPEALS
FOR THE SIXTH CIRCUIT

FILED

NOV - 7 2000

LEONARD GREEN, Clerk

BOYD E. GRAVES,

Plaintiff-Appellant,

v.

WILLIAM S. COHEN, et al.,

Defendants-Appellees.

ORDER

NOT RECOMMENDED FOR FULL-TEXT PUBLICATION

Sixth Circuit Rule 28(g) limits citation to specific situations. Please see Rule 28(g) before citing in a proceeding in a court in the Sixth Circuit. If cited, a copy must be served on other parties and the Court.
This notice is to be prominently displayed if this decision is reproduced.

Before: MERRITT, WELLFORD, and SILER, Circuit Judges.

Boyd E. Graves, a pro se Ohio resident, appeals a district court order dismissing his civil rights complaint filed pursuant to 42 U.S.C. § 2000e (Title VII), 29 U.S.C. § 791 (the Rehabilitation Act of 1973), and 42 U.S.C. § 1985. This case has been referred to a panel of the court pursuant to Rule 34(j)(1), Rules of the Sixth Circuit. Upon examination, this panel unanimously agrees that oral argument is not needed. Fed. R. App. P. 34(a).

Seeking monetary relief, Graves sued multiple federal officials in the United States Pentagon contending that the federal government developed and purposely proliferated a synthetic biological agent on the American population. The agent Graves identified subsequently became known as acquired immune deficiency syndrome (AIDS). Graves states that as a result of the government's action, he acquired AIDS in the 1980's.

On September 16, 1999, the district court dismissed Graves's allegations as frivolous except for Graves's vague claim of employment discrimination. The district court subsequently allocated Graves thirty days to file an amended complaint in order to set forth sufficient facts to give the defendants notice of his employment discrimination claims. After numerous extensions of time and orders to amend his complaint, Graves declined to amend the complaint. As Graves failed to comply with the court's numerous orders, the complaint was dismissed.

In his timely appeal, Graves argues that the district court erred in denying his request to represent a class of AIDS sufferers, that he was entitled to appointed counsel, that his complaint did not warrant injunctive restrictions, that his exhibits satisfied the federal rules, and that the district court erred by not allowing service of process on the defendants.

We review the district court's order de novo. *See McGore v. Wrigglesworth*, 114 F.3d 601, 604 (6th Cir. 1997).

The district court properly dismissed the complaint as frivolous. Graves alleged that the government engaged in an ultra secret biological program which created the AIDS virus. This virus was then released into the American population by the Pentagon. A complaint is frivolous if it lacks an arguable basis either in law or in fact. *See Neitzke v. Williams*, 490 U.S. 319, 325 (1989). Such claims describe fantastic or delusional scenarios. *Id.* at 328. As Graves's claim concerning the AIDS virus being injected into the American population by the Pentagon has no basis in law or in fact, the district court did not err in dismissing Graves's allegation concerning the virus as frivolous.

In dismissing a case for want of prosecution, a district court should consider whether a plaintiff's claim is fictitious, whether the delay indicates dilatory conduct, the claimant's responsibility for the delay, whether lesser sanctions are available, and whether the delay prejudiced the defendant. Absent specific notice that dismissal is contemplated or a record showing a party's bad faith, dismissal is improper. *See Vinci v. Consolidated Rail Corp.*, 927 F.2d 287, 287-88 (6th Cir. 1991); *Harris v. Callwood*, 844 F.2d 1254, 1256 (6th Cir. 1988). Dismissal pursuant to Fed. R. Civ. P. 41(b) is proper where there exists a clear record of delay or contumacious conduct. *See Little v. Yeutter*, 984 F.2d 160, 162 (6th Cir. 1993).

The facts establish that in a period of a year, the district court constantly ordered Graves to amend his complaint to state a valid employment discrimination claim. However, Graves chose to ignore the multiple orders of the district court. Given Graves's dilatory conduct, the district court did not err in dismissing Graves's complaint. Finally, given the facts of this case, we conclude Graves's arguments on appeal are meritless.

Accordingly, we affirm the district court's order. Rule 34(j)(2)(C), Rules of the Sixth Circuit.

ENTERED BY ORDER OF THE COURT

Clerk

No. 99-4476

UNITED STATES COURT OF APPEALS

FOR THE SIXTH CIRCUIT

BOYD E. GRAVES,)

)

Plaintiff-appellant,)

)

v.) **Circuit Judges:**

)

WILLIAM S. COHEN, et. al.) **MERRITT, WELLFORD AND SILER**

)

Defendant-appellees.)

Appeal from the Northern District of Ohio

Case No. 98CV2209

MOTION FOR RECONSIDERATION OF THE COURT'S ORDER OF NOVEMBER 7, 2000 BASED ON JUDICIAL ERROR

Plaintiff-appellant, Boyd E. Graves, pro se, requests reconsideration of the Court's Order of November 7, 2000 based on judicial error; failure to properly review the docket below.

1. The November 7, 2000 Order does not review the complaint filed on September 28, 1998 for adherence to FRCP 4(a). The complaint filed on October 28, 1998 meets the federal rules for

sufficiency of fact and should have been serve upon the named defendants.

2. The November 7, 2000 Order makes no reference to the "additional facts" provided to the Court below on November 2, 1998. Immediately following receipt of the district court's Order of October 28, 1998, plaintiff-appellant submitted an AMENDED COMPLAINT in the form of a motion for reconsideration. SEE Docket Entry #3 (11/2/98). The 'amended complaint' of November 2, 1998 and the original complaint of September 28, 1998 should have been served on the named defendants in the interest of swift, speedy resolution of federal civil litigation. The precedence of this Court allows for a pro se motion for reconsideration to be liberally construed as an amended complaint, prior to service of process of the defendants.

3. Following the exhaustion of the federal EEO administrative procedure, plaintiff-appellant was notified by the federal agencies of his Constitutional right to file a civil procedure in federal court. The complaint filed on September 28, 1998 was timely in accordance with the final notices issued by the agencies. The federal defendants were well aware of the ensuing federal action with respect to the "well-established" allegations of employment discrimination in violation of the Rehabilitation Act of 1973, the Americans with Disabilities Act, and the various civil rights acts. This matter had unduly persisted in the federal EEO process since 1995, prior to the filing in September, 1998.

4. At all times relevant to the processing of this matter, plaintiff-appellant dutifully met his burden of providing "sufficient facts" to the court below. Plaintiff-appellant asks for reconsideration of his September 28, 1998 complaint and November 2, 1998 motion for reconsideration as sufficient to meet the requirements of the federal rule. For the convenience of the Court, plaintiff-appellant is resubmitting Exhibit Two (attached). The docket of the case below does contain an amended complaint and sufficient facts in accordance with the Federal Rules of Civil Procedure in which service of process should have been effected without additional submissions to the docket record.

5. Pursuant to 28 USC 1915(e), because plaintiff-appellant was proceeding in forma pauperis, the court below reviewed plaintiff-appellant's allegations of AIDS bioengineering for frivol-

ity (unsupported allegations). The court's Order of November 7, 2000 excludes the fact that the plaintiff-appellant's allegations were supported by substantial, credible documentation and testimony, including the Congressional Record of the United States. SEE Exhibit One (9/28/98). The court's Order of November 7, 2000 is in error because plaintiff-appellant provided the court below substantial and sufficient facts, documentation and evidence in support of his allegations. Plaintiff-appellant's allegations of AIDS bioengineering are thus, not frivolous (unsupported) pursuant to Section 1915(e) because they are supported by sufficient facts and substantial, credible documentation, including a government flowchart revealing the "research logic" of the development of AIDS. Plaintiff-appellant's allegations are not 'bare face' and as such easily clear the bar of frivolity of Section 1915(e). Additionally, this Court was provided a May 15, 2000 letter from Dr. Victoria Cargill, Medical Officer for the Office of AIDS Research of the National Institutes of Health **confirming** the existence of the secret federal virus development program.

6. In light of the substantial supporting documentation of a U.S. Pentagon program, the "Special Virus" ("AIDS") , the Court's Order of November 7, 2000 is in serious error to the detriment of the U.S. Constitution and the class of American people who have been killed, injured and maimed by another U.S. Pentagon project of human subject experimentation. This Court must reconsider that victims of this U.S. Pentagon project are entitled to "judicial accountability" and an answer to the coplaint. This Court must reconsider its errant Order of November 7, 2000 based on the plain error identified in this motion and the will and intent of the American people. The docket does contain a legitimate complaint supported by sufficient facts and extensive documentation for each allegation. Clearly, the dismissal scheme of Section 1915(e) does not apply in this instant action. The district court is not free to "set aside" evidence to reach a finding of frivolity under Section 1915(e). Poor plaintiffs, proceeding pro se, should not be held to extra- judicial procedures and extraordinary submissions prior to service of process, in contradiction to the Sixth Circuit and Supreme Court precedence. Equal access to justice allows for plaintiff-appellant's November 2, 1998 motion for reconsideration to be construed as an amended complaint. It is also clear the named federal

EEO defendants were well aware of the anticipated federal action when they granted plaintiff-appellant the right to file suit within 30 days of the completion of the EEO administrative process.

WHEREFORE; plaintiff-appellant, Boyd E. Graves, prays the Court will reconsider its Order of November 7, 2000 and allow for service of process in a fair-minded district court of the United States of America. Poor plaintiffs should not be required to provide extraordinary proof of their claims prior to service of process pursuant to 28 USC 1915(e). Plaintiff-appellant's complaint of September 28, 1998 is consistent with the requirement standards established by the U.S. Supreme Court and the U.S. Constitution. See **Haines V. Kerner**, 404 U.S. 519, 520-21 (1972).

Respectfully submitted,

Dated: November 11, 2000

Boyd E. Graves, pro se

3844 E. 140th Street

Cleveland, OH 44128

216-561-1967

CERTIFICATE OF SERVICE

I, Boyd E. Graves, do hereby certify that I served a copy of the foregoing motion for reconsideration on Lisa Hammond Johnson, U.S. Attorney, sent via first class mail this ____day of November, 2000, postage prepaid.

1

____

BOYD E. GRAVES

No. 99-4476

UNITED STATES COURT OF APPEALS
FOR THE SIXTH CIRCUIT

FILED
JAN 12 2001
LEONARD GREEN, Clerk

BOYD E. GRAVES, Plaintiff-Appellant, v. WILLIAM S. COHEN, et al., Defendants-Appellees.	)	ORDER

Before: MERRITT, WELLFORD, and SILER, Circuit Judges.

Boyd E. Graves petitions for rehearing of this court's order of November 7, 2000, which upheld the district court's judgment dismissing Graves's civil rights action. After careful consideration, the panel concludes that it acted under no misapprehension of law or fact in issuing this order. Fed. R. App. P. 40(a).

Accordingly, the petition for rehearing is denied.

ENTERED BY ORDER OF THE COURT

Leonard Green
Clerk

No. 99-4476

UNITED STATES COURT OF APPEALS
FOR THE SIXTH CIRCUIT

BOYD E. GRAVES,
Plaintiff-Appellant

V.

WILLIAM S. COHEN, et al,
Defendants-Appellees

PETITION FOR REHEARING EN BANC

The Appellant's exhibit (a DOD flowchart) requires a full review of the Court as to the issue of AIDS bioengineering. The appellant's exhibit alone passes the bar of frivolity identified by the lower court pursuant to Section 1915(e).

It is a travesty to the class of victims of this federal virus development program that this Court seeks to set aside the indisputable evidence of the true laboratory origin of the AIDS pandemic. The Special Virus program must be reviewed at some level of the trilateral government structure as compelled by the U. S. Constitution.

In any petition to the Supreme Court the class will seek to present the following question as representative of error of this Circuit:

Is the 1971 flowchart of the Special Virus program of the United States of America "sufficient evidence" to preclude a finding of frivolity pursuant to Section 1915(e) of the United States Code?

It is the class' position the flowchart and progress reports are conclusive proof of a federal program seeking to make a contagious cancer that selectively kills, by depleting the immune system. The Court en banc can not overlook the scientific significance of the completeness of the adjudication of the true origin of AIDS.

The complaint filed on September 28, 1998 requires a response as demanded by the United States Constitution consistent with **Haines v. Kerner, 404 U.S. 519 - 521 (1972).** It is contrary to the social order of the common good of the United States to foster stealth programs that kill. The attached 5/15/00 letter from the National Institutes of Health confirms the reality of the program. The attached letter from the U.S. Attorney identifies the office of the true defendant best evidenced in P.L.91-213, signed March 16, 1970 by former President Richard M. Nixon. May God hath mercy on the United States, my country.

Respectfully submitted,

Boyd E. Graves, pro se, in forma pauperis
P.O. Box 332
Abilene, KS 67410
785-263-1871

Dated: January 22, 2001

January 27, 2001

The Honorable Lesley Brooks Wells, Judge
United States District Court
Northern District of Ohio (Eastern Division)
201 Superior Avenue, N.E.
United States Court House
Cleveland, OH 44114

RE: 98-2209
Graves v. Cohen (Rumsfeld)
Sixth Circuit No. 99-4476

Dear Judge Wells:

Yesterday I received a mandate from the Sixth Circuit that confirms your October 28, 1998 finding that I am delusional as to the true origin of AIDS. Your belief that AIDS has an African etiology is seriously challenged by the presentation to you of the 1969 sworn Pentagon testimony, and the 1971 flowchart of the Special Virus Program of the United States.

On Wednesday I will conduct a briefing to Dr. Eric Goosby of the Surgeon General's office on my research and evidence. In April I will represent the United States on this issue in a delegation to South Africa according to Dr. James Hall.

You have effectively weakened the Constitution to the detriment of poor people living with HIV/AIDS. Yet at the same time you cavort within the Black community as if it were part of your heritage.

Two hundred Black people die every hour and you continue to wear your blindfold in which to actively repress justice and access to the court.

I will seek further review of the federal program, the Special Virus. It is unfortunate that your lifetime appointment allows you the discretion to set aside evidence for your own judicial activism. Ultimately, the 1971 flowchart will prove itself to be one of the greatest document finds in the history of the world.

Of course, for you and the Gods of the Sixth Circuit, the flowchart should probably just be recycled as were the first seven progress reports of the program. Although you have succeeded in casting me as "crazy", it is odd that my evidence continues to be confirmed at the highest eschelons of our democracy. This case best exemplifies the facade of the judicial structure of the United States. The world will not allow your finding of frivolity to hold. This case does require an extraordinary review at the district court level.

In that same vein, your dismissal pursuant to Section 1915(e) was available to you because I have been reduced to abject poverty at the hands of this federal program. In addition to all other avenues that I will pursue to seek further accountability of this secret program, I will resubmit the complaint of September 28, 1998 with the appropriate filing fee.

People with HIV/AIDS are entitled to an ANSWER and an account of the greatest genocidal secret in the history of the world.

Respectfully,

Boyd E. Graves
P.O. Box 332
Abilene, KS 67410
785-263-1871

enc: A) 5/15/00 ltr, Cargill, Victoria, M.D., Medical Officer, Office of AIDS Research National Institutes of Health, Bethesda, MD.

B) The 1971 Flow Chart ("RESEARCH LOGIC") of the Special Virus program of the United States.

cc: The Honorable George W. Bush, President
The United States of America

The Honorable Kofi Annan, Secretary General
The United Nations

The Honorable David Satcher, Surgeon General
The United States of America

The Honorable Tommy Thompson, Secretary
Health and Human Services

The Honorable Trent Lott, Majority Leader
The United States Senate

The Honorable Dennis Hastert, Speaker
The United States House of Representatives

Admiral Craig Quigley, Spokesperson
The United States Pentagon

Leonard Green, Clerk
United States Court of Appeals for the Sixth Circuit

Geri Smith, Clerk
U.S. District Court

Dr. Leonard Horowitz

Dr. Robert Strecker

No.______________

In The

SUPREME COURT OF THE UNITED STATES

October Term, 2000

BOYD E. GRAVES, et. al.

Petitioners,

vs.

THE PRESIDENT OF THE UNITED STATES OF AMERICA, et. al.,

Respondents.

On Petition for Writ of Certiorari to the United States Court
of Appeals for the Sixth Circuit

PETITION FOR WRIT OF CERTIORARI

Boyd E. Graves, pro se
PO BOX 332
ABILENE, KS 67410

Petitioner

Ted Olsen, Solicitor General
Room 5614
Department of Justice
950 Pennsylvania Avenue, N.W.
Washington, DC 20530-0001

Counsel for Respondents

TABLE OF CONTENTS

U.S. Law

PL91-213, "Population Growth and The American Future", Nixon, R. President of the United States of America, March 16, 1970
The Genocide Convention Implementation Act of 1987 (The Proxmire Act)
National Security Council Memorandum #46, Brezinski, Z., "Africa and the US Black Movement", March 17, 1978
U.S. House Resolution 15090

U.S. Supreme Court Case Law

Boag v. MacDougall, 454 U.S. 364, 365 (1982) (per curiam)
Bounds v. Smith 430 U.S. 817 (1977)
Haines V. Kerner, 404 U.S. 519 (1972)
Nietzke v. Williams, 490 U.S. 319 (1989)
Procunier v. Martinez, 416 U.S. 396 (1974)

U.S. Sixth Circuit Case Law

Blalock, 983 F.2d 1353 (6th Cir. 1993)
Knop v. Johnson, 977 F.2d 996, 1003 (6th Cir. 1992)
Lawler V. Marshall, 898 F.2d 1196 (6th Cir.)
Lillard V. Shelby County Board of Education, 76 F.3d 716
McGore v. Wrigglesworth, 114 F.3d 601. (1997)

Medical and Scientific Citations

"Viral Infections in Man Associated With Acquired Immunological Deficiency States", Merigan, T.C. et. al., Federation Proceedings, Vol. 30, No. 6, November-December, 1971

Sonigo, P. et. al. Cell 1985 Aug;42(1):369-382, "Nucleotide Sequence of the Visna Lentivirus; Relationship to the AIDS Virus"

Proceedings of the United States of America, National Academy of Sciences, Volume 83 pp. 4007 - 4011, (June 1986) ("evolutionary relationship between AIDS and VISNA")

Proceedings of the United States of America, National Academy of Sciences, Volume 92, pp. 3283 - 3287, (April 11, 1995). (submitted by Carlton Gajdusek, M.D.)

Gallo, R. et. al., Science, January 1985, pp173 -177 ("HIV equals VISNA")

FEMS MICROBIOL LETT 128 (1995) 63 -68, "Mycoplasmas Regulate HIV"

Garth Nicolson, "Mycoplasmal Infections in Chronic Illnesses: HIV/AIDS", Medical Sentinel 1999; 4:172-176

Science News, August 1996, J. Craig Venter, Jr. (Human Genome Project)

Special Virus. program of the United States of America, pp. 1- 61, Progress Report #8, (1971).

Gallo, R.C, "Reverse Transcriptase in Type C Virus Particles of Human Origin", pg. 335 (1971) PR#8, U.S. Special Virus program (**AIDS DEVELOPMENT PAPER #650**).

Montagnier, L., & Gallo, R.C. et. al. "Human T-Cell Leukemia Lymphoma Virus" (Cold Spring Harbor Laboratory, Cold Spring Harbor, NY, 1984).

Shepherd, M.C. 1954, "The Recovery of Pleuropneumonialike organisms from Negro Men With and Without Nongonococcal Urethritis" Amer. J. Syphilis Gonorrhea Venereal Dis. 38:113-124.

Graves, B.E., "MYCOPLASMA VISNA: The Scientific Name of AIDS." Day 97: WORLD WAR AIDS, March 7, 2001, personal communication.

Graves, B.E., "STATE ORIGIN: The Evidence of the Laboratory Birth of AIDS, Zygote Media Network, P.O BOX 332 Abilene, KS 67410

U.S. Patents

Patent No. 4647773 (AIDS)

Patent No. 5112756 (VISNA)

Patent No. 5871745 (Multiple Sclerosis)

APPENDICES

APPENDIX A
-Order-Sixth Circuit-DENIAL OF PETITION FOR REHEARING

APPENDIX B
-PETITION FOR REHEARING

APPENDIX C
-Order-Sixth Circuit-Affirming AIDS BIOENGINEERING- FRIVOLOUS

APPENDIX D
-Final Order-District Court

APPENDIX E
-Order-District Court-AIDS BIOENGINEERING-FRIVOLOUS

APPENDIX F
-Rule 32, Diagram and Exhibit List and Exhibits

OVERVIEW OF THE PETITION

Boyd E. Graves petitions the high Court to seek resolution of a burning social issue not previously addressed by the Court. Since the onset of AIDS in 1979, less and less scrutiny has been focused on the genesis of the pandemic, in which to (allegedly) concentrate on education, treatment and prevention. After receiving an HIV positive diagnosis in 1992, petitioner began a relentless pursuit to find the truth.

Exhibit One submitted to the District Court on September 28, 1998 attached to the complaint reveals the U.S. Pentagon informed the U.S. Congress on June 9, 1969 they could make AIDS. See, U.S. House Resolution 15090, Part VI, page 129. ("APPENDIX F", EXHIBIT 4). Petitioner subsequently found the "research logic" Flow Chart of a secret U. S. virus development program. ("APPENDIX F", EXHIBIT 3). At issue before the Court is whether petitioner's exhibits in support of his allegations of AIDS bioengineering can be set aside in which to reach a finding of frivolity pursuant to 28 USC 1915(e)? Additionally, are people with HIV/AIDS entitled by law to an apology and review of the secret federal program?

It is the petitioner's position the "equal protection/equal access" clauses of the U.S. Constitution preclude the district court from "setting aside" evidence in which to reach a finding of frivolity. Additionally, petitioner believes the subsequent presentation to the courts below of the 1971 "research logic" Flow Chart proves absolutely the "design", "intent" and "purpose" of a federal virus development program of the United States. The AIDS virus is a hybrid ("synthetically recombined") virus because petitioner and the class can prove the virus has been specifically developed in accordance with the Flow Chart and fifteen progress reports of the secret federal program. See, APPENDIX F, EXHIBIT 9, see also, e.g. Sonigo, P. et. al. Cell 1985 Aug;42(1):369-382, "Nucleotide Sequence of the Visna Lentivirus; Relationship to the AIDS Virus", "Viral Infections in Man Associated With Acquired Immunological Deficiency States", Merigan, T.C. et. al., Federation Proceedings, Vol. 30, No. 6, November-December, 1971. According to all scientific criteria, the human wasting virus ("AIDS") is identical to the animal-wasting virus ("VISNA"). See, Proceedings of the United States of America, National Academy of Sciences, Vol. 83, pp. 4007 - 4011, June 1986.

Petitioner's allegations of AIDS bioengineering submitted to the district court on September 28, 1998 were supported by official U.S. documents and pass any frivolity bar in the civilized world but for the judicial activism affirmed by the lower court. This case must be heard to ensure our democracy is truly at the heart and center of our Constitutional form of government.

According to the experts, Dr. Len Horowitz, et. al., it is this petitioner, Boyd E. Graves, who best represents the voice of the class of persons whose Constitutional rights of equal protection and equal access (to the courts) would be eliminated. To not hear the sound legal principles of any issue is an egregious act, to set aside a properly brought case "sufficiently complete" as frivolous is contrary to the public good and the public conscience in every case. Those principles must be enhanced to ensure poor people with HIV/AIDS have a Constitutionally protected right of equal access to the federal courts and an equal apprehension of the law.

This high Court once again has the opportunity to be the "gate keeper of reason" for ALL the people of the United States. This Court has an opportunity to deride the social lunacy pervasive throughout former President Nixon's PL91-213 and put and end to stealth population culling in the name of population stabilization. ("APPENDIX F", EXHIBIT 6 & 7).

In this extraordinary request to be heard, we meet our extraordinary proofs requirement with **the**

"research logic" Flow Chart of the 1962 - 1978 federal virus development program, the Special Virus. ("APPENDIX F", EXHIBIT 3).

The U.S. Special Virus program spent $550 million dollars ("APPENDIX F", EXHIBIT 5) to make a virus and has never accounted for the program to the American people **until now**. Although we have been indignantly turned away by the lower courts, we believe the public significance of this issue exceeds the prior frivolity decisions in such a fashion that this Court **must** exercise its proper judicial discretion and hear this case.

QUESTIONS PRESENTED FOR REVIEW

DO THE RIGHTS OF POOR PEOPLE WITH HIV/AIDS NEED TO BE STRENGTHENED REGARDING ACCESS TO THE COURTS AND EQUAL PROTECTION IN RELATION TO 28 USC 1915(e)(B)(i)?

IS THE 1972 FLOW CHART DOCUMENT DEFINITIVE PROOF OF A FEDERAL VIRUS PROGRAM?

JURISDICTION IN SUPPORT OF PETITION FOR WRIT

This petition for writ of certiorari is brought pursuant to the Court's jurisdiction incumbent under 28 USC 1254(1). The Sixth Circuit Court of Appeals chose to affirm a lower court ruling of AIDS bioengineering frivolity on November 7, 2000. ("APPENDIX C"). After a timely filed motion for reconsideration (liberally construed as a petition for rehearing), ("APPENDIX B"), the Sixth Circuit denied the petition for rehearing on January 12, 2001. ("APPENDIX A"). Pursuant to Rule 32, petitioner submitted his diagrams and exhibits to the clerk for inspection on March 27, 2001. ("APPENDIX F").

LIST OF PARTIES

1. Boyd E. Graves, Petitioner (Class Representative)
P.O.Box 332
Abilene, KS 67410

2. The President of the United States (UNITED STATES RESPONDENTS)
c/o **Ted Olson, Solicitor General** (designee)
ROOM 5614
Department of Justice
950 Pennsylvania Avenue, N.W.
Washington, D.C. 20530-0001

3. John Wodatch, Chief (in the care of the Solicitor General)
U.S Department of Justice
Disability Rights Division
1425 New York Ave., N.W.
ROOM 4039
Washington, D.C. 20005

4. Lawrence Roffee (in the care of the Solicitor General)
James Raggio
David Capozzi
U.S. Architectural and Transportation Barriers Compliance Board
1331 F Street, N.W.
Suite 1000
Washington, D.C. 20004-1000

5. Robert Gallo, COO (in the care of the Solicitor General)
Institute of Human Virology (Special AIDS Virus Developer)
725 W. Lombard Street
Baltimore, MD 21201

6. Alan Rabson, Deputy Director (in the care of the Solicitor General)
National Cancer Institute (Special AIDS Virus Developer)
31 Center Drive, Bldge 31
ROOM 11A48
Bethesda, MD 20892

7. Garth Nicolson, COO, CEO (in the care of the Solicitor General)
The Institute for Molecular Medicine (Special AIDS Virus Developer)
15162 Triton Lane
Huntington Beach, CA 92649-1401

8. Peter Duesberg (in the care of the Solicitor General)
7835 Faust Ave (Special AIDS Virus Developer)
Canoga Park, CA 91304

9. Shi-Ching-Lo (in the care of the Solicitor General)
U.S. Army Institute of Pathology (patent holder-pathogenic mycoplasma)
Fort Detrick
Frederick, MD 21702

10. (yet named federal defendants)

STATEMENT OF THE CASE

Petitioner, Boyd E. Graves, like other members of the class, took an AIDS test (in 1992) and learned he was HIV Positive. After experiencing HIV discrimination from U.S Disability agencies (identified), petitioner brought suit (9/28/98) in the U.S. District Court, Northern Division (EASTERN), Cleveland, Ohio. The Case was assigned to Judge Lesley Brooks Wells, who on October 28, 1998 ruled the plaintiff's allegations as to AIDS bioengineering "frivolous" pursuant **explicitly** to 28 USC 1915(e)(B)(i). ("APPENDIX E"). In doing so, **the district court "set aside" petitioner's exhibit evidence in support of his AIDS bioengineering allegations.** id. The district court also ruled the complaint and exhibits filed on September 28, 1998 did not state any factual basis for the "HIV employment discrimination" claim. id. See also, 2/22/00, APPELLEE'S BRIEF filed by Lisa Hammond Johnson, Esq., United States Attorney.

It is the petitioner's position each and every allegation of the complaint filed on September 28, 1998 is thoroughly supported by substantial, credible evidence and federal Marshall service of process should have ensued. The petitioner does not believe the federal rules, the United States Code nor any of the Circuit Courts of Appeal allow U.S. documents to be "set aside" in which to trivialize any issue. There has been a grave miscarriage of justice which without review by this Court, there will never be a fair review of the U.S. federal virus program, the Special Virus.

Petitioner and the class have been turned away from inherent rights longwithstanding in the Constitution and the Courts . Petitioner believes this representative case is the "best eyeglass" to truly reflect a fair and just society, 'gilbraltered' on truth and fact and equal justice under law for ALL people; especially those, **purposefully targeted** with the U.S. Special AIDS virus. We have found the wellspring of the genesis of AIDS; it is a program of our nation state. It is us. **People with HIV/ AIDS are entitled under every provision of the U.S. Constitution to an immediate apology and an immediate review of this federal program.**

In 1971, the United States knew the animal disease, VISNA, had never before been seen in

human disease. See, pg. 39, Progress Report #8, U.S. Special (AIDS) Virus program. In 1995, it is our own U.S. National Academy of Sciences who concludes the "identical sequencing" in AIDS and VISNA. See, Proceedings of the United States of America, National Academy of Sciences, Volume 92, pp. 3283 - 3287, (April 11, 1995). The Academy's conclusions are consistent with that of Dr. Robert Gallo's, the "co-discoverer" of AIDS. See, (Gallo, R. et. al., Science, January 1985, pp173 -177. But for the U.S. Special Virus program, VISNA would not have an evolutionary relationship to AIDS. In light of the scientific experiments and medical papers incumbent in the fifteen (yearly) progress reports of this federal program, the Special Virus, ("APPENDIX F", EXHIBITS 3, 5, & 9) the people should not have to wait another day for this Court to utilize the fail safe mechanism of descretionary review. **Visna is natural only in AIDS.** Visna is a transmissible spongiform encephalapathy similar to Mad Cow in cattle and variant Creutzfeldt Jacob disease in humans. They are made transmissible ("infectious") by tiny prokaryotes called "mycoplasmas". Mycoplasmas have no cell wall, and their "involvement" in the pathogenesis of AIDS is a well documented undisputed scientific fact. See, FEMS MICROBIOL LETT 128 (1995) 63 -68, "Mycoplasmas Regulate HIV", See, also Garth Nicolson, et. al. "Mycoplasmal Infections in Chronic Illnesses: HIV/AIDS", Medical Sentinel 1999; 4:172-176 . The record reveals: MYCOPLASMA VISNA was first tested on mammals in SOUTH AFRICA in 1915, in Montana in 1923 and the full HIV/AIDS precursor in 1932 in Iceland. According to the government narrative (pages 1 - 60 of Progress Report #8 (1971), U.S. Special Virus program), it does appear the Flow Chart and the 15 progress reports of the U.S. Special Virus program will definitively explain the "cross species transmission" experiments to realize this animal disease in the human genome. See, Graves, B.E., "MCOPLASMA VISNA: The Scientific Name of AIDS", DAY 97: WORLD WAR AIDS, March 7, 2001, see also, Graves, B.E., "STATE ORIGIN: The Evidence of the Laboratory Birth of AIDS", Zygote Media Network, Abilene, KS (Summer 2001). **It must be argued that 'even in 1915', Visna appears to have a laboratory origin.**

Your rise to the forefront of this overwhelming evidence will allow for the review process to begin the lengthy purification of the darkest chapter in human history. May God have mercy on the United States.

REASONS FOR GRANTING THIS PETITION

This petition for writ of certiorari is submitted with a direct focus on the national importance of the definitive proof of the genesis of AIDS. Without certiorari, the United States Constitution will continue to "ring hollow" for poor people with HIV/AIDS. But for the judicial activism that has "set aside" significant case law and evidence, the American people, particularly those with HIV/AIDS, would have received an apology and "make whole" relief in some yet described manner some time prior, long ago. Poor people (with HIV/AIDS) are full citizen Americans and as such are entitled to the full foundation protections outlined in the U.S. Constitution. In light of the overwhelming evidence of a federal virus development program (the Special Virus), justice requires this petition for certiorari be granted.

The dismissal scheme incumbent in the United States code Section 1915(e)(B)(i) was not intended to provide a "sidestep mechanism" to allow district court judges the opportunity to unceremoniously dismiss cases of significant issues, simply because they have been brought by poor people seeking monetary relief. See, Sixth Circuit, 11/7/00 (ELECTION DAY) ORDER setting aside the Class' Flow Chart as frivolous. ("APPENDIX A").The lower courts refused to rule on the supportive direct evidence (1971 Flow Chart, 1969 Congressional Testimony, President Nixon on signing PL91-213, Nixon on July 18, 1969, Brezinski on March 17, 1978, the program's $550 million budget, Cargill, M.D. "official verification" 5/15/2000, etc.) in which to cast the petitioner as "delusional". ("APPENDIX E").

Petitioner has no delusion over the direct evidence of AIDS bioengineering. Petitioner's complaint on behalf of the Class of HIV/AIDS victims filed on September 28, 1998 was not reviewed by the Sixth Circuit for compliance with Rule 4(a) of the Federal Rules of Civil Procedure in accordance with Sixth Circuit law. See, McGore v. Wrigglesworth, 114 F.3d 601. (1997). The complaint filed on September 28, 1998 met and exceeded the federal rules for "sufficiency"., Reason finds it is the district court standing in the way of the Constitution for this class of citizens.

Additionally, to ensure equal access for poor, unrepresented citizens, there is normally built-in "wide latitude" in the statutory framework for pro se litigants. Boag v. MacDougall. **If there had been a technical defect in the original complaint, it was immediately cured on November 2, 1998 when petitioner submitted a "Motion for Reconsideration" (liberally construed as an AMENDED COMPLAINT).** Any de novo McGore review would conclude the complaint filed on September 28, 1998 should have been served by the Federal Marshall on the United States defendants. It does appear that this petitioner has sufficiently outlined a legitimate necessity for the granting of the writ of certiorari. Although this Court has previously addressed the sufficiency of poor people's complaints, the issue has never arisen with respect to a class of HIV/AIDS Americans. Without this Court's direct intervention and leadership, poor people with this federal virus (HIV/AIDS) will be forever relegated to despondency, despair and ultimate hopelessness. This direction is deleterious to Constitutional foundations. **Surely this high Court will require an accounting of this federal program.** A review by this Court is the one reassuring way citizens with HIV/AIDS will know they have standing and relevance in this resourceful American society. As it is now, you have secretly maintained a federal virus program to the detriment of the American people and the United States Constitution. Once we learned of the program and sought redress, our direct evidence is "set aside", we are billed as 'delusional', and the perpetrators continue to circumvent the U. S. Constitution by devising better and better biological ethnic killing weapons. See, Science News, August 1996, Craig Venter (Human Genome Project , owner of the African American ethnic gene patent, CCR5, the entry-way for the U.S. Special AIDS virus. Poor people with HIV/AIDS are entitled to better than this from a government which now conclusively wears two faces and is clearly not what it seems. We have relentlessly presented our evidence to both the Executive and Legislative branches of government to no avail. We look to this Court to begin the process for resolution of the last American secret of the 20th Century. We have found the wellspring of the genesis of AIDS; it is a federal program called the Special virus. ("APPENDIX F, EXHIBIT 3, 5, & 9). Each and every HIV/AIDS victim is innocent and entitled to an apology and in nearly every case, "make whole" relief. The direct evidence leaves but one conclusion, this case must be heard, The GENOCIDE CONVENTION IMPLEMENTATION ACT OF 1987 ("The PROXMIRE ACT" of 1987).

THE EVIDENCE OF THE LABORATORY BIRTH OF AIDS

On March 27, 2001, petitioner, Boyd E. Graves (Class Representative) submitted a ten exhibit package to the Clerk of the Supreme Court in accordance with Rule 32 of this Court. ("APPENDIX F"). Since March 27, 2001, the United States Supreme Court has possessed a copy of the "research logic" of the U.S. Special Virus program without comment. According to a growing number of doctors and scientists, (Lee, Loren, Horowitz, Cantwell, Halstead, Boyle, Nicolson, etc.) this Flow Chart document is the **"missing link"** in resolving the 'why' of the "sudden emergence" of the AIDS virus in the late 1970's and early 1980's. See, www.boydgraves.com/comments.

The AIDS virus is the candidate virus sought by the United States under a federal program entitled the Special Virus. See, pp. 1- 61, Progress Report #8, (1971).The U.S. Special Virus program sought to mix "Visna" and other animal virus particles in human tissue cultures to see if the animal virus particles would trigger "cell death" (neoplastic transformation) in human cells. id. According to page 39 of the 1971 report, the animal disease, "Visna" had never before been seen in human disease until HIV/AIDS.

The Flow Chart and fifteen progress reports of this federal virus development program provide the only plausible explanation for the "sudden explosion appearance" of Visna in the human population. According to all scientific criteria, HIV has significant fragments of the genetic sequence of Visna. See, (Robert Gallo, et. al), Science, January 1985, pp. 173 - 7.

In April 1984, Dr. Robert Gallo announced the discovery of the mystery 'wasting' illness. It was a retrovirus, however, a cursory review inside the '71 report reveals that Dr. Gallo's 1971 experiment paper in the Special Virus program is identical to his 1984 announcement. See, Gallo, R.C, "Reverse Transcriptase in Type C Virus Particles of Human Origin", pg. 335 (1971) PR#8, U.S. Special Virus program (AIDS DEVELOPMENT PAPER #650). Without oversight of this Court, Dr. Gallo, et. al. will never be made to explain. Equally Dr. Gallo should be made to explain the 'patent relationship' between his 1984 U.S. AIDS patent (No.: 464773)and U.S. Visna and Multiple Sclerosis patents (Nos.: 5112756 and 5871745). Additionally, the Court must note an "intellectual property relationship" between the infectious agent of AIDS and the infectious agent of Chronic Fatigue Immune Deficiency Syndrome a.k.a. Fibromyalgia Encepholapathy.

In 1971, John B. Moloney, (deceased) Associate Science Director for the National Cancer Institute announced the candidate virus would be a "leukemia/lymphoma" virus. See, Progress Report #8 at 2 (1971). Prior to its current name change to 'hiv', AIDS was originally called a 'leukemia/lymphoma' virus. See, Montagnier, L., & Gallo, R.C. et. al. "Human T-Cell Leukemia Lymphoma Virus" (Cold Spring Harbor Laboratory, Cold Spring Harbor, NY, 1984).

We have successfully navigated a maze and matrix inside an illusionary federal system of equal protection and equal access. At the ('never before reached') finish line, we find an AIDS ORIGIN ODYSSEY heavy laden with curtains and false propaganda. We must collectively pay attention to the curtains and the men behind them, the U.S. Constitution compels it. Those curtains are the Flow Chart, experiments, contracts, scientific and medical papers of the five hundred fifty million dollars stealth federal program that made a "death immune system virus" to the detriment of humanity and in absolute contradiction of the United States Constitution and the free will of the American people.

Will a non-invasive diagnostic oral swab procedure to test citizens for antibodies to Visna be necessary, before people with HIV/AIDS receive an apology and (similar to) Tuskegee "make whole" relief?

In other words, the total history of the development of AIDS 'appears to run parallel' with the lengthy 20th Century experiments dealing with African American sperm transmissions. See, Shepherd, M.C. 1954, "The Recovery of Pleuropneumonialike organisms from Negro Men With and Without Nongonococcal Urethritis" Amer. J. Syphilis Gonorrhea Venereal Dis. 38:113-124. Visna may in actuality turn out to be a sheep 'syphilis' mycoplasma in the human system. In any oral presentation on this issue, we will further assert the United States' top AIDS doctors, Dr. David Satcher, Dr. Eric Goosby, Dr. Victoria Cargill had never heard of the evolutionary relationship between HIV and its next of kin, Visna, prior to this information being provided to them by the class of HIV/AIDS citizens.

LEGAL ARGUMENTS SUPPORTING THIS PETITION

Petitioner believes the complaint filed on September 28, 1998 was legally sufficient as to named and yet-named defendants as to allow federal Marshall service of process. Boag. Petitioner believes the lower courts "purposefully" set aside well established case law both of this Court and the Sixth Circuit. See, e.g Boag v. MacDougall, 454 U.S. 364, 365 (1982) (per curiam); Haines V. Kerner, 404 U.S. 519, 20 -1, (1972) or more specifically, Lillard V. Shelby County Board of Education, 76 F.3d 716, 724 (quoting Mertik v. Blalock, 983 F.2d 1353, 67-8 (6th Cir. 1993), see also, Knop v. Johnson, 977 F.2d 996, 1003 (6th Cir. 1992), cert denied, 507 U.S. 973 (1993), Bounds v. Smith 430 U.S. 817, 823 (1977) quoting Ross v. Moffitt, 417 U.S. 600, 611 (1974) (interchanging "poor plaintiffs with AIDS" for 'prisoners').

Petitioner believes a complaint can not be dismissed as frivolous pursuant to 28 USC 1915(e)(B)(i) if it is supported by credible evidence, no matter how outlandish the allegations of the complaint may be. The intent of all the Circuits and Congress has never been to use the dismissal scheme affiliated with the filings of poor people as a judicial mechanism to preclude access and equal protection for poor people. See, McGore V. Wrigglesworth, 114 F.3d 601. In this regard, I represent a class of people. A group of citizens who have been inflicted with a "SYNTHETIC BIOLOGICAL AGENT THAT DEPLETES OUR IMMUNE SYSTEM SO AS TO ALLOW FOR THE ONSET OF INFECTIOUS DISEASE". ("APPENDIX F" (EXHIBIT 4, page 129). Poor people with HIV/AIDS are entitled to seek to exercise their full Constitutional rights in light of the definitive proof of a stealth federal virus program "persisting with fervor" from 1962 to 1978. The 1972 Flow Chart submitted to this court for inspection ("APPENDIX F", EXHIBIT 3) on March 27, 2001 has been authenticated as an official document by Dr. Victoria Cargill ("APPENDIX F", EXHIBIT 2), Medical Officer, Office of AIDS Research, National Institutes of Health. According to Dr. Cargill, Dr. Alan Rabson, Deputy Director of the National Cancer Institute, was forwarded her 5/15/2000 letter of "official verification" of the 1971 Flow Chart and progress reports of the U.S. Special Virus program.

It is clear the lower courts have acted in contradiction to the ideals and principles of equal justice and equal access under law when it comes to preserving, strengthening and protecting the rights of the victims of this federal experimentation program. This petition graphically depicts grave misapprehension of core Constitutional principles and American values. **At every juncture it is the United States who is guilty of contumacious conduct**. Specifically, the underlying federal AIDS discrimination issues were 'culminating' in the 9/28/98 complaint. The federal disability agency defendants were "well aware" of the prior lame three-year federal EEO process. The complaint filed on 9/28/98 was in direct response to final notices issued by each of the agencies. It is ludicrous and preposterous to portray the complaint filed on September 28, 1998 as insufficient to the federal rules of civil procedure or the United States Code. The petitioner's pleadings were sufficient to continue the Constitutionally protected judicial process of SERVICE OF PROCESS and an ANSWER.

Petitioner believes this individual case meets every requirement for review in accordance with Rule 10 of this Court. This petition for writ should be granted to prove poor people with HIV/AIDS are equally protected once wrongdoing is uncovered and redress is sought. **We have been erroneously vilified and denied basic Constitutional rights to effect a population stabilization scheme designed solely for economic and resource greed.** See, Africa and the U.S. Black Movement, Zbigniev Brezinski, 3/17/78, ("APPENDIX F", EXHIBIT 10). The holdings of Denton is such that a finding of frivolity can not be maintained. This petition for writ of certiorari should be granted based on the number of extraordinary and significant legal and social issues involving the life and liberty of poor people with HIV/AIDS.

We have pulled back the AIDS curtain and pursuant to Rule 32 submitted many official U.S. documents for your inspection. These extraordinary documents support this class of plaintiffs' allegations of AIDS bioengineering. The United States must be compelled to answer.

"By what natural mechanism does the Icelandic Sheep disease, Visna, appear in the genetic sequence of HIV"?

We believe we thoroughly demonstrate the existence of the "compelling reason" standards necessary for the Court to exercise its rare discretionary jurisdiction.

Finally, pursuant to Procunier v. Martinez, 416 U.S. 396, 419 (1974), it is exceedingly clear the dismissal scheme was invalid and service of process should immediately proceed. Petitioner's claims of AIDS bioengineering have an arguable basis in both law and fact. His claims are based on e.g., PL 91-213 ("APPENDIX F", EXHIBIT 6) and U.S. House Resolution 15090 ("APPENDIX F", EXHIBIT 4). Petitioner's claims of AIDS bioengineering are supported by official U.S. documents and as such, Nietzke v. Williams, 490 U.S. 319, 325 (1989) was further misapplied by the lower court. In light of the voluminous federal documents, the Sixth Circuit's November 7, 2000 Order is grossly inconsistent with well established law best exemplified by Lawler V. Marshall, 898 F.2d 1196 (6th Cir.).

CONCLUSION

On behalf of the American victims of this Special AIDS virus program, we believe we are entitled by the United States Constitution to an apology, "make whole" relief and a standard of care reflective of the spirit and soul of the "true" founding intentions of which we incorporate. We will not stray from our guiding direction of Equal Justice Under Law. In light of the many people who have signed our petition seeking review of this federal virus program, www.boydgraves.com/petition, it does appear 'the whole world is watching'.

Petitioner, Boyd E. Graves, respectfully prays his efforts have genuinely garnered the interests of the Court on behalf of the innocent victims of the U.S. Special AIDS Virus program.

Respectfully submitted,

Boyd E. Graves, in forma pauperis and pro se
PETITIONER (CLASS REPRESENTATIVE)
P.O. BOX 332
ABILENE, KS 67410
785-263-1871

CERTIFICATE OF SERVICE

I, Boyd E. Graves, petitioner and class representative, do hereby certify with my signature below that I served a complete copy of this petition for writ of certiorari on the Solicitor General and named U.S. Attorney, Lisa Hammond Johnson on behalf of George W. Bush, President of the United States of America and the named and yet named federal respondents, sent this ____day of April, 2001 via first class mail postage prepaid to:

George W. Bush, President, The United States of America
John Wodatch, Chief
U.S. Department of Justice
Lawrence Roffee, Executive Director
James Raggio
David Capozzi
USATBCB
Robert Gallo, M.D.
Alan Rabson, M.D.
Garth Nicolson, Ph.D
Peter Duesberg, M.D.
Shi-Ching Lo, M.D.
(and yet-named federal respondents)

c/o Barbara Underwood, The Solicitor General of the United States of America (designee)
Room 5614
Department of Justice
950 Pennsylvania Ave, N.W.
Washington, D.C. 20530-0001
Lisa Hammond Johnson, Esq.
United States Attorney
Northern District of Ohio
1800 Bank One Center
600 Superior Avenue, East
Cleveland, OH 44114-2654

____

Boyd E. Graves, in forma pauperis and pro se
PETITIONER (CLASS REPRESENTATIVE)
P.O. BOX 332
ABILENE, KS 67410
785-263-1871

IN THE SUPREME COURT OF THE UNITED STATES

GRAVES, BOYD E., ET AL.
Petitioner

vs.

THE PRESIDENT OF THE U.S.

No. 00-9587

WAIVER

The Government hereby waives its right to file a response to the petition in this case, unless requested to do so by the court.

BARBARA D. UNDERWOOD
Acting Solicitor General
Counsel of Record

May 10, 2001

cc: BOYD E, GRAVES
P.O. BOX 332
ABILENE, KS 67410

5/16
TO: Dr. David Satcher, Surgeon General
the Solicitor General has waived opposition
to our petition, even though AIDS poses
a death sentence and is in essence,
"capital punishment" via a 'state-managed'
federal virus program, the Special Virus!

Boyd E. Graves

No.: OO-9587

In The

SUPREME COURT OF THE UNITED STATES

October Term, 2000

BOYD E. GRAVES, et. al.

Petitioners,

vs.

THE PRESIDENT OF THE UNITED STATES OF AMERICA, et. al.,

Respondents.

On Petition for Writ of Certiorari to the United States Court of Appeals for the Sixth Circuit

SUPPLEMENTAL BRIEF IN OPPOSITION TO THE UNITED STATES WAIVER APPLICATION OF MAY 10, 2001

Boyd E. Graves, pro se
P.O. BOX 332
Abilene, KS 67410
Petitioner

Barbara D. Underwood
Acting Solicitor General
Room 5614
Department of Justice
950 Pennsylvania Avenue, N.W.
Washington, DC 20530-0001

Counsel for Respondents

Petitioner, Boyd E. Graves, hereby files a supplemental brief to compel the government to file a reply brief in the above listed case.

On or about April 11, 2001, petitioner filed a petition for writ of certiorari seeking review of a federal Flow Chart as definitive proof of a federal virus development program. On May 10, 2001, the United States responded by entering the appearance of the acting Solicitor General and seeking to waive its right to file a reply brief (attached). Pursuant to the Rules of the Court, petitioner argues a reply brief is mandatory and the Court should move with deliberate speed to require (Order) the government to respond to the substantial and credible allegations of AIDS bioengineering and state management. f the 1971 Flow Chart is definitive proof of a federal virus development program, then petitioner and the class are technically under a "death sentence" pursuant to Rule 14.1(a). Between 1962 and 1978, the United States, with "design", "purpose" and "intent", sought to create new human viruses. The evidence firmly shows the AIDS virus is a chimera-hybrid (animal/human) virus likely originating under the auspices of the U.S. Special Virus program. Petitioner's class is afflicted with a "man made wasting syndrome" called "VISNA". See, Petition for Writ, April 11, 2001. Indeed, because of the inclusion of the man made Visna in the genetic sequencing of HIV, (See, page 39, progress report #8 (1971, U.S. Special Virus program) each and every member of this class has been relegated to a cruel, stealth form of 'capital punishment' and in essence, Constitutional rape.

Petitioner, Boyd E. Graves, respectfully prays the Court will ORDER the United States government to reply to the direct evidence of a federal virus program, e.g., the 1971 Flow Chart and 15 progress reports (1963-1978) and to the posted comments of the science and medicine experts. See, http://www.boydgraves.com/comments. To allow the Solicitor General the opportunity to 'side-step' the Constitution is contrary to every established principle of our democracy. The American people want nature to determine the finish line for the human race, not man. The United States Constitution compels our government to address the secret virus development program.

Respectfully submitted,

May 17, 2001

Boyd E. Graves, pro se
P.O. Box 332
Abilene, KS 67410

PHASE V

CLINICAL TRIALS

www.boydgraves.com

PHASE V: CLINICAL TRAILS

Signature Petition for Immediate Review

Additional Research Materials

Order Form

PETITION FOR IMMEDIATE REVIEW

On March 3, 2000, Boyd E. Graves, J.D. hand delivered over three thousand signature petitions to the U.S. Congress calling for the immediate review of the United States's Special Virus Program (1962-1978). Over a year later the United States' has yet to take any action toward reviewing this secret tax funded virus development program which produced 60,000 liters of synthetic immune suppressing virus in 1977. This petition continues gaining international momentum, bipartisan support and will continue collecting signatures until the federal virus development program of the United States, the "Special Virus" is reviewed. The petition is circulated on the web, via email, and postal mail. Dr. Graves will hand deliver these signature petitions to the United States Congress. Please circulate this petition far and wide.

"I the undersigned hereby add my signature to the People's petition seeking demanding the immediate independent review of the U.S. Special Virus Program and the Evidence of Laboratory Birth of the AIDS virus."

Name	Address	City	State	Zip	Telephone	Email
Name	Address	City	State	Zip	Telephone	Email
Name	Address	City	State	Zip	Telephone	Email
Name	Address	City	State	Zip	Telephone	Email
Name	Address	City	State	Zip	Telephone	Email
Name	Address	City	State	Zip	Telephone	Email
Name	Address	City	State	Zip	Telephone	Email
Name	Address	City	State	Zip	Telephone	Email
Name	Address	City	State	Zip	Telephone	Email
Name	Address	City	State	Zip	Telephone	Email

Sign Petition On-Line: www.boydgraves.com/petition

Send Petitions to: N.O.A.H. - PO Box 332 - Abilene, KS - 67410 1-800-257-9387

Place
Stamp
Here

National Organization for the Advancement of Humanity
c/o Zygote Media Networks
PO Box 332
Abilene, KS 67410

ATT: SVCP REVIEW PETITIONS ENCLOSED

PETITION FOR IMMEDIATE REVIEW

On March 3, 2000, Boyd E. Graves, J.D. hand delivered over three thousand signature petitions to the U.S. Congress calling for the immediate review of the United States's Special Virus Program (1962-1978). Over a year later the United States' has yet to take any action toward reviewing this secret tax funded virus development program which produced 60,000 liters of synthetic immune suppressing virus in 1977. This petition continues gaining international momentum, bipartisan support and will continue collecting signatures until the federal virus development program of the United States, the "Special Virus" is reviewed. The petition is circulated on the web, via email, and postal mail. Dr. Graves will hand deliver these signature petitions to the United States Congress. Please circulate this petition far and wide.

"I the undersigned hereby add my signature to the People's petition seeking demanding the immediate independent review of the U.S. Special Virus Program and the Evidence of Laboratory Birth of the AIDS virus."

Name	Address	City	State	Zip	Telephone	Email
Name	Address	City	State	Zip	Telephone	Email
Name	Address	City	State	Zip	Telephone	Email
Name	Address	City	State	Zip	Telephone	Email
Name	Address	City	State	Zip	Telephone	Email
Name	Address	City	State	Zip	Telephone	Email
Name	Address	City	State	Zip	Telephone	Email
Name	Address	City	State	Zip	Telephone	Email
Name	Address	City	State	Zip	Telephone	Email
Name	Address	City	State	Zip	Telephone	Email

Sign Petition On-Line: www.boydgraves.com/petition

Send Petitions to: N.O.A.H. - PO Box 332 - Abilene, KS - 67410 1-800-257-9387

Place
Stamp
Here

National Organization for the Advancement of Humanity
c/o Zygote Media Networks
PO Box 332
Abilene, KS 67410

ATT: SVCP REVIEW PETITIONS ENCLOSED

VIDEOS

"AIDS: MIRACLE IN CANADA" The most important video in history is now available to the public. Relive the AIDS origin lecture presented by American Boyd Graves, J.D. His presentation in the 'airplane rotunda' of the prestigious Western Reserve Historical Society is wrought with scientific and legal precision as to the absolute laboratory birth of AIDS. 128 mins. VHS **$24.95 + $4 S/H**

"AIDS: MADE IN THE U.S.A." On October 29, 1999, American Boyd Ed Graves, J.D. established the secret 1971 U.S. AIDS Flow Chart as the greatest document find of the century. IN this review of his landmark legal case for a Global AIDS Apology, "AIDS: MADE IN THE U.S.A." is one of the most informative videos on the true laboratory birth of AIDS. 160 min. VHS **$24.95 + $4 S/H**

BOOKS

"STATE ORIGIN: The Evidence of The Laboratory Birth of AIDS" State Origin is a gripping review of the direct government *evidence of the laboratory birth of the AIDS* pandemic. Author Boyd Ed Graves J.D. has spent the last ten years tracking the origins of AIDS through a densely-covered propaganda trail of the highest deception. The shocking conclusions of Graves' research led him to file the landmark federal case **Graves vs. The President of the United States, United States Supreme Court Case No. 00-9587. The AIDS holocaust has an activist, it is Boyd E. Graves and this is his story. $29.95 + $4 S/H**

"THE 100 DAY WAR" On December 1, 2000, Boyd E. Graves, J.D. travelled to America's heartland and declared the event, WORLD WAR AIDS DAY. Dr. Graves initiated a one man educational offensive on AIDS. The World War AIDS collection was written in Kansas and included essays, 27 radio interviews, lectures, and heated dialogue with the U.S. Surgeon General, AIDS 'codiscoverer' Robert Gallo, and many more. Dr. Graves' efforts to awaken the world to the diabolical intent of the mostly secret U.S. Special Virus program has brought us closer to taking AIDS apart by using the blue print and experiments they used to make it. The 100 Day War on AIDS, helped raise public awareness and signature petitions in order to never again allow the greatest democracy in the world the opportunity to masquerade. (Now In Press)

U.S. SPECIAL VIRUS PROGRESS REPORTS (1962-1978): According to National Cancer Institute librarian, Judy Grossberg, the agency archives only three of the 15 known annual Special Virus progress reports. Dr. Graves' research has uncovered more of the annual reports than the government admits even exists: The experiments inside the progress reports are part of the serious allegations now made against the government before U.S. Supreme Court. The 1971 Flow Chart discovered in 1999 coordinates every contract and experiment found inside these secret pages and has been called the "greatest document find of the century". The 'missing reports' Dr. Graves research has helped find are close to 500 pages each and are offered for $100 per report, plus shipping and handling charges.

April 2001

ONE MINUTE . . .

Graves Moving U.S. To Review Special Virus Program

"One Minute . . ." with Boyd E. Graves, J.D. is the only newsletter dedicated to the people's independent review of the U.S. Special Virus Cancer Program (1962-1978). The monthly "One Minute..." also keeps readers up to date on the latest developments in this landmark legal case for justice. Monthly issues include breaking news, essays, events, photographs, special announcements in a quick informative publication. "One Minute . . ." can be delivered via U.S. Post Office, or directly to your email inbox.

Individual subscription rates are $100 annually, or just $25.00 billed quarterly. For corporate subscriptions or sponsorship information, please call 1-800-257-9387.

E-BOOK SUPPLEMENTAL - For readers who purchased the e-book, the supplemental package includes the same shocking collection of nonbound documents included in the printed version. This document collection is offered for an additional fee of $10. E-book customers can alternatively choose to take $5 off the purchase price of the printed book simply by including the e-book receipt with your paid order.

"Be Something Special"

Be an affiliate. Webmasters, entrepreneurs, veterans, students, medical historians, doctors, vets, and innocent victims of the U.S. Special Virus have linked for education and news for months. Now the e-book encourages people to link for profit. Every link established by e-book resellers, can earn you up to 50% from e-books sold by your links. Every resold book and signed petition generated by your efforts help Dr. Graves close the darkest chapter of the 21st Century. For more information please send email to **affiliate@boydgraves.com**

"With respect for the living and dignity for the dead, We demand Equal Justice Under Law for every innocent victim of the U.S. Special Virus development program."

Boyd E. Graves addressing the public on the steps of the U.S. Supreme Court April 11, 2001

*Credit card charges will appear on your bill as: Zygote Media Networks.

**Personal checks for books, videos, subscriptions, and reports may be made to: Zygote Media Networks

***Donations may be made payable to: National Organization for the Advancement of Humanity

Product Title		Price	QTY.	Total
AIDS: Miracle In CANADA	VHS Video 1999	$24.95	__	_____
AIDS: Made in the U.S.A.	VHS Video 1999	$24.95	__	_____
STATE ORIGIN: The Evidence of the Laboratory Birth of AIDS		$29.95	__	_____
1971 AIDS FLOWCHART	11 x 32 poster	$10.00	__	_____
"One Minute . . ."	Annual Subscription	$50.00	__	_____
E-book Supplemental	Document Collection	$10.00	__	_____
Special Virus Program Progress Report #8 1971	NIH/NCI	$50.00	__	_____
Special Virus Program Progress Report #9 1972	NIH/NCI	$50.00	__	_____
Special Virus Program Progress Report #10 1974	NIH/NCI	$50.00	__	_____
Special Virus Program Progress Report #14 1977	NIH/NCI	$50.00	__	_____
Special Virus Program Progress Report #15 1978	NIH/NCI	$50.00	__	_____
Donation to Graves vs President of the United States		$_____	__	_____
Shipping & Handling		$4.00 per item	__	_____
KS residents add .06 % sales tax		$_____	__	_____
		TOTAL	__	_____

Bill & Ship To:

Name:______________________ MC / VISA: ______________________

Address:______________________ cc:______ ______ ______ ______

City:______________________ exp: ____________

State:______ Country: ______________

Zip: ______________

Phone: ______________ e:mail: ______________

CHECK/MONEY ORDER# __________

MAIL ORDER INFORMATION
N.O.A.H.- c/o Zygote Media

PO Box 332 - Abilene, KS 67410 Toll Free: 1-800-257-9387

CHECK BY FAX: 1-785-263-1568 (Paste check on separate sheet)

www.boydgraves.com

Place
Stamp
Here

National Organization for the Advancement of Humanity
c/o Zygote Media Networks
PO Box 332
Abilene, KS 67410

ATT: BOOK FULFILLMENT

Seeking Justice: The sun rises over the federal virus flow chart, exposing the global AIDS genocide plot and offering the world renewed hope.

"The U.S. Special Virus program spent $550 million dollars ("APPENDIX F", EXHIBIT 5) to make a virus and has never accounted for the program to the American people until now." -

Boyd E. Graves, J.D.

The Man Who Solved AIDS

On August 21st, 1999 in Ganonoque, Canada, Boyd E. Graves, B.S., J.D. unveiled the U.S. government's secret 1971 AIDS Flow Chart and experiments , stunning the international medical and scientific communities and forever answering the true origin of AIDS.

The Flow Chart is the "Research Logic Flow" of an ultra-secret federal program entitled the "Special Virus" (1962-1978). Since August 1999, the Flow Chart continues to receive intense scrutiny with many scientists and medical doctors now going on record as confirming the diabolical nature of the "Special Virus" program. The "Special Virus" is the designer product of a century long hunt for a contagious cancer that will selectively kill. The "necessity" for the creation and deployment of AIDS, et.al. is fully outlined in U.S. population control policy decisions including "National Security Study Memorandum 200 (NSSM-200) 1974, written and presented by Henry Kissinger at the mega-conference on population control held in Bucharest,Romania. In 1969 Public-Law 91-213 signed by Richard M. Nixon legalized 'population stabilization' and the use of the government's new Special Virus, AIDS.

In 1992 Dr. Graves began his research on the global AIDS pandemic in earnest. Today he is serving his second year as Director-AIDS CONCERNS for the international medical research foundation, Common Cause, headquartered in Ontario, Canada. His AIDS research led him to the Special Virus Flow Chart discovery in 1999. The government's disturbing "research logic" led Dr. Graves to his AIDS activism and role as Lead Plaintiff for Global AIDS Apology.

Dr. Graves civil rights, human rights, and AIDS activism has received critical acclaim from around the world. He has challenged our nation to an irreversible review of its' secret virus program, while challenging society to awaken to the reality of the greatest human holocaust in mankind's history.

Dr. Graves hand delivered the people's appeal brief to the United States Supreme Court on April 11, 2001. The People of the world await the court's decision.

Printed in the United States
28910LVS00001B/23

9 780970 773517